MEDICINE OF PLACE

Patterns of Nature and Psyche

in the Wildflowers of Cascadia

Julia M. Brayshaw

ALCHEMIA PUBLISHING

© 2007 by Julia M. Brayshaw.

Medicine of Place, Wildflowers of Cascadia Card Deck © 2007 by Karen Lohmann.

All rights reserved. No part of this publication may be reproduced, stored in a retrieval system, or transmitted, in any form or by any means, electronic, mechanical, photocopying, recording, or otherwise, without the prior written permission of the author.

All of the information in this book—including descriptions of traditional and contemporary applications of plants—is offered solely for educational and inspirational purposes. In no way is it intended for the diagnosis or treatment of any condition, and in no way does it replace the counsel of a trusted and qualified health practitioner. Each individual is unique and complex, and the author supports self-empowerment, careful research, and active collaboration in making informed, responsible choices appropriate to the uniqueness of each situation.

Cover Photo: Columbia Hills Natural Area Preserve: Kevin Head.
All wildflower photos by Kevin Head.
Cover and text design by Liina Koivula.

ISBN-10: 0-9798674-4-4
ISBN-13: 978-0-9798674-4-6

First printing 2007

For additional copies or inquiries, please contact the author:

Julia Brayshaw
424 38th Ave. NE
Olympia, WA 98506
juliamb33@hotmail.com

or visit:
www.medicineofplace.com

*Dedicated with love to my mother,
Emily Nalence Buxbaum,
who, in her 90s, still counsels: "Follow your dreams."
...and to Kevin, the "man in the kayak".*

What we love asks us to be its ambassador in the court of the human world.

—Caroline Casey

Contents

Acknowledgments   ix

Definitions   xiii

Alphabetical Plant List   xvi

Part I: Introduction

   Chapter One: Reweaving our World   1

   Chapter Two: Suggestions for Using this Book and Wildflower Card Deck   12

      Table 1   17

      Table 2   19

   Chapter Three: The Medicine of Flowers   20

      Table 3   29

Part II: The Wildflowers of Cascadia:

   Thirty-three Monograms

Grass Widow	34
Salmonberry	38
Swamp Lantern	43
Red-flowering Currant	48
Western Trillium	53
Columbia Desert Parsley	59
Poet's Shooting Star	63
Common Camas	67
Bitterroot	73
Bleeding Heart	78

Calypso Orchid	83
Shining Oregon Grape	88
Nootka Rose	93
Oregon Iris	98
Rock Penstemon	103
Upland Larkspur	110
Simpson's Hedgehog Cactus	117
Large-leaved Lupine	122
Sagebrush Mariposa	127
Great Blazing Star	132
Mountain Lady's Slipper	135
Purple Virgin's Bower	142
Columbia Lewisia	147
Jessica's Stickseed	152
Wyethia	156
Tweedy's Lewisia	161
Beargrass	165
Fireweed	172
Common Paintbrush	177
Shrubby Cinquefoil	183
River Beauty	186
Scotch Bluebell	191
Explorer's Gentian	197
Glossary	201
References	207
The Author and Artist	212

Author's Acknowledgments: "This *is* the Garden"

In the biggest sense, Medicine of Place has been dreamed into being through the wondrous Cascadia bioregion, nourished by an extensive community, and called forth and guided by the wildflowers themselves. It owes its existence to a life-long imperative to drop everything for events of beauty in nature. Countless hours devoted to the yearly unfolding of wildflowers have finally preempted concerns that there are more important things to do. Collaborating on Medicine of Place has allowed me to whole-heartedly value my relationship with flowering plants, to follow my curiosity about our encounters with the natural world, and to attend to the universal patterns that shape both inner and outer nature.

Without the help of many others, staying true to authentic self would be impossible. I mention just a few who represent the indispensable many —you know who you are!

The Flower Essence Society bestowed multiple gifts, helping my path to open before me, naming the truth that had been with me always: "Flowers are medicine for the soul."

I am grateful to many courageous clients who have allowed me to learn that through careful attention, outer forms reveal the inner essence.

I deeply acknowledge the First Nations People who belong to Cascadia, on whose land I live, work, play, and benefit. Reverent stewards of this land for millennia, they are true guardians of the medicine of place. May we respectfully allow them to lead us back to right relationship.

On Candlemas of this year David Pond beautifully articulated the bigger truth that now demands revelation. Through this articulation, my life, my "job description", this writing all became coherent, and I recognized my own awakening within a greater one. Rosie Finn distilled David's words into one "mantra": "This is the Garden". Finally exposed as ugly rumor is the notion that Earth is anything less than sacred. (Does this make a difference in how we relate to the Earth? asks David.) I am filled with gratitude for a network of luminaries unfolding this truth—indigenous people, dancers, artists, cooks, green healers, body workers, deep ecologists, reverent farmers, music makers—all who feed the senses, who direct us to nature's splendid table. They reveal the grace of dwelling on Earth in the body's luminous architecture.

Other invaluable helpers include Carol Trasatto, Elise Krohn, Shelley Rudeen, Victoria Tennant, and Elizabeth Markell. Thank you to Jordan Taylor for having the same "job description", for presenting me with a clear reflection of its complete worthiness. I am gratefully indebted to Jeanne Lohmann whose skillful editing was actually a much needed course in writing. Debi Bodett's beautiful eye for color and space brought the wildflower cards to life. Thank you, Debi, both for your wonderful graphic design work and for your guidance and encouragement that helped us first put pen and pastel to paper! And big thanks to Liina Koivula whose thoughtful design work transformed pages of words to a lovely finished product.

Special gratitude goes to artist Karen Lohmann for dreaming the same dream, for the ability to move from vision to manifestation, and for the life and beauty of the pastel portraits without which Medicine of Place would not be. Nor would it be without Kevin Head, whose photographs have added a whole dimension to this work—your reverence for each flower shines through them! Thank you, Kevin, for many treks to the wilderness, for your wonderful naturalist skills, and for the encouragement, support, and belief in me that many times saved this project from the recycle bin. You are golden—a true Blazing Star! Most especially I am grateful to my daughters, Jamie and Dayna, whose heart-opening presence in my life, like the flowers, is a constant affirmation of life's beauty and goodness.

Julia Brayshaw
Spring, 2007
Olympia, WA

Artist's Acknowledgments

There are so many people who have helped support this project, and perhaps I have left folks out, but please know you are each honored in the work. I wish to thank: Julia and Kevin for sharing and holding this vision, working through the process and for creating such poetic guidance with wisdom of the plants and land; my mother, Jeanne Lohmann for all of her kind, sweet support, encouragement and editing help!; my husband, Joe Tougas, for his even, deep thinking, unwavering help and love; my children Sam and Ramona, for their passion for knowledge, all they teach me, their humor and kindness; Morgan, Chie and Grandson Dean Teruki for the joy they bring; my brothers; my women's group for years of buttressing and fun; Marilyn Frasca for inspiration and vision; Debi Bodett for her spirit and designing skill; my interfaith community, CIC; Jo Curtz and the Chanting group; Elise Krohn, Elizabeth Markell, Laurie McEvers, Carol Trasatto, Judy Manley, Shelley Kirk-Rudeen, Victoria Tennant, Isabel Keefe, Robin Landsong, Candace Vogler and cousin Dotty for the many ways they have helped to shape and hone this project. Many thanks also to Providence SoundHomeCare and Hospice for opportunities to know life's tapestry more fully; and to Richard Katz and Patricia Kaminski of the Flower Essence Society for deep guidance and inspiration along this floral path. Praise to the bees, wildflowers, soil and mysteries which feed us all!

Karen Lohmann
Summer, 2007
Olympia, WA

DEFINITIONS

medicine:

    If I find a green meadow
    splashed with daisies and sit down
    beside a clear-running brook, I
    have found medicine.

    —Deepak Chopra

    Lord, the air smells good today, straight from the
    mysteries within the inner courts of God.
    A grace, like new clothes thrown
    across the garden, free medicine for everybody!

    —Rumi

place:

    Where ever you are is called HERE.
    And you must treat it as a powerful stranger.
    Must ask permission to know it and be known.
    The forest breathes. Listen. It answers,
    I have made this place around you.

If you leave it, you may come back again, saying HERE.
No two trees are the same to Raven.
No two branches are the same to wren.
...The forest knows
Where you are. You must let it find you.

__David Wagoner

medicine of place:

We need the tonic of wilderness, to wade sometimes in marshes where the bittern and the meadow-hen lurk, and hear the booming of the snipe; to smell the whispering sedge where only some wilder and more solitary fowl builds her nest, and the mink crawls with its belly close to the ground.

__Henry David Thoreau

Cascadia:

a bioregion often referred to as the Pacific Northwest. Named for its "abundance of cascading waterfalls", this bioregion "can be defined as the watersheds of rivers that flow into the Pacific Ocean through Turtle Island's temperate rain forest zone (Turtle Island is the native term used for the continent commonly referred to as North America)....The physical geography that is Cascadia extends from southern Alaska to northern California, and from the Pacific coast to the continental divide, taking in Washington, most of

Oregon, Idaho and British Columbia, and parts of Alberta, Montana, and California. This area is distinguished by zones of similar climate, vegetation, and animal species."[1]

bioregion:

"a life place, a location that is specific to a culture, to distinct flora and fauna, and to the soils, weather, and topology that are characteristic of that place. Though a bioregion is a geographical location, it is also a frame of mind, a way of thinking that includes recognizing one's sense of place, both historically through culture, and through physical interactions with that place....Bioregions are...living systems in which all life-forms are interconnected and interrelated."[2] A bioregion is "...a whole system comprised of a set of diverse, integrated natural sub-systems...run by ecological laws with which humans...must work in cooperation if there is to be a sustainable future."[3]

---

1   Mocniak, Jeff. *Cascadia; getting to know your bioregion.*
2   *Ibid.*
3   Berg, Peter. (quoted in Mocniak)

Alphabetical Plant List:

BOTANICAL NAMES OF PLANTS REFERRED TO IN THE TEXT.
(Bold entries indicate plants featured in monograms and include page number.)

Alpine Buckwheat	*Eriogonum pyrolifolium*	
Arnica	*Arnica*	
Arrowleaf Balsamroot	*Balsamorhiza sagittata*	
Aster	*Aster*	
Avalanche Lily	*Erythronium montanum*	
Balsamroot	*Balsamorhiza*	
Beargrass	***Xerophyllum tenax***	p. 165
Bitterroot	***Lewisia rediviva***	p. 73
Bleeding Heart	***Dicentra formosa***	p. 78
Blue-Eyed Mary	*Collinsia*	
Calypso Orchid	***Calypso bulbosa***	p. 83
Cascades Penstemon	*Penstemon serrulatus*	
Cascade Oregon Grape	*Mahonia nervosa*	
Coltsfoot	*Petasites palmatus*	
Columbia Desert Parsley	***Lomatium columbianum***	p. 59
Columbia Lewisia	***Lewisia columbiana***	p. 147
Common Camas	***Camassia quamash***	p. 67
Common Paintbrush	***Castilleja miniata***	p. 177
Creeping Oregon Grape	*Mahonia repens*	
Davidson's Penstemon	*Penstemon davidsonii*	
Devil's Club	*Oplopanax horridus*	
Evening Primrose	*Oenothera*	
Explorer's Gentian	***Gentiana calycosa***	p. 197
Fireweed	***Chamerion angustifolium***	p. 172
Foxglove	*Digitalis purpurea*	

Gairdner's Penstemon	*Penstemon gairdneri*	
Grass Widow	***Olsynium douglasii***	p. 34
Gray's Desert Parsley	*Lomatium grayi*	
Great Blazing Star	***Mentzelia laevicaulis***	p. 132
Green Bog Orchid	*Habenaria hyperborea*	
Gold Star	*Crocidium multicaule*	
Heather	*Phyllodoce*	
Horsetail	*Equisetum*	
Huckleberry	*Vaccinium*	
Jessica's Stickseed	***Hackelia micrantha***	p. 152
Large-Leaved Lupine	***Lupinus polyphyllus***	p. 122
Lupine	*Lupinus*	
Mimulus	*Mimulus*	
Mountain Lady's Slipper	***Cypripedium montanum***	p. 135
Mountain Pride	*Penstemon newberryi*	
Mountain Sweet Cicely	*Osmorhiza chilensis*	
Nootka Rose	***Rosa nutkana***	p. 93
Oregon Iris	***Iris tenax***	p. 98
Osoberry	*Oemleria cerasiformis*	
Paintbrush	*Castilleja*	
Pearly Everlasting	*Anaphalis margaritacea*	
Pedicularis	*Pedicularis*	
Pink Fawn Lily	*Erythronium revolutum*	
Pink Heather	*Phyllodoce empetriformis*	
Poet's Shooting Star	***Dodecatheon poeticum***	p. 63
Prairie Lupine	*Lupinus lepidus*	
Prairie Star	*Lithophragma parviflorum*	
Puget Balsamroot	*Balsamorhiza deltoidea*	
Purple Virgin's Bower	***Clematis columbiana***	p. 142
Red-flowering Currant	***Ribes sanguineum***	p. 48
River Beauty	***Chamerion latifolium***	p. 186

Rock Penstemon	*Penstemon rupicola*	p. 103
Rosy Spiraea	*Spiraea splendens*	
Round-leaved Bog Orchid	*Habenaria orbiculata*	
Sagebrush	*Artemisia*	
Sagebrush Mariposa	*Calochortus macrocarpus*	p. 127
Salmonberry	*Rubus spectabilis*	p. 38
Salt and Pepper Lomatium	*Lomatium piperi*	
Saxifrage	*Saxifraga*	
Scotch Bluebell	*Campanula rotundifolia*	p. 191
Shining Oregon Grape	*Mahonia aquifolium*	p. 88
Shrubby Cinquefoil	*Potentilla fruticosa*	p. 183
Simpson's Hedgehog Cactus	*Pediocactus simpsonii*	p. 117
Stinging Nettles	*Urtica dioica*	
Stonecrop	*Sedum*	
Swamp Lantern	*Lysichiton americanus*	p. 43
Thyme Buckwheat	*Eriogonum thymoides*	
Tweedy's Lewisia	*Lewisia tweedyi*	p. 161
Upland Larkspur	*Delphinium nuttallianum*	p. 110
Valerian	*Valeriana sitchensis*	
Waterleaf	*Hydrophyllum fendleri*	
Western Trillium	*Trillium ovatum*	p. 53
Wyethia	*Wyethia amplexicaulis*	p. 156
Yellow Bells	*Fritillaria pudica*	

Part I: Introduction

The visible world was made to correspond to the world invisible and there is nothing in this world but is a symbol of something in that world.

—Abu Hamid Muhammad al-Ghazzali

Chapter One: Reweaving our World

Nature is not only what is visible to the eye—it shows the inner images of the soul—the images on the back side of the eyes.

—Edvard Munch

Our psyche is set up in accord with the structure of the universe, and what happens in the macrocosm likewise happens in the infinitesimal and most subjective reaches of the psyche.

—Carl Gustav Jung

Medicine is everywhere in wild places. Nature extends her sensuous hand, offering solace and inspiration. Each of her interwoven facets is a unique expression of a great and beautiful generative dance.

Medicine of Place blends visual art and narrative to highlight the special medicine offered through wildflowers. It celebrates our magnetic, mysterious relationship to them. The plants that bear flowers are linked to our very existence. Their seeds, fruits, and nectars supply critical nutrients, without which we and many other life forms could never have developed.[1] We need only imagine a world without flowers to realize that these plants that have learned to blossom are indispensable also to the nourishment of our souls.

Wildflowers express the creative and regenerative power of the Earth. Their beauty proclaims the sanctity of the land. We are blissfully helpless to describe what truly happens when we walk among milky statues of Beargrass, find magenta pools of Grass Widows by winter oaks, or sit among the jewel-like colors of a mountain meadow. Our experience tells us there is indeed strong medicine here. Metaphor and analogy, ancient wisdom, modern physics, and depth psychology help walk us to the edge of this mystery.

Each native plant is like a beautiful song. This song belongs to its whole species; at the same time it follows the themes and patterns of the plant's landscape. It harmonizes with the music of other beings around it, and together they offer the big song of their place. The more deeply we listen to native plants and creatures the more we become entrained to the resonance and rhythms of their surroundings. They assist us to be in harmony with the world around us, and therefore to come into inner harmony as well.

In this way, native plants and creatures are carriers of medicine. Medicine is what heals, that is, makes whole. We experience wholeness when our relationships, both inner and outer, are in harmony. Through participating in mutually enhancing, life-giving connections we contribute to the systems within and outside of ourselves; we experience our place in the web of life. Wild beings can induct us into the music of an ecosystem. They carry the medicine of place.

A slightly different way to talk about this is to say that native plants hold the stories of their habitats and regions, and that these are healing stories, stories of wholeness. Each story is the plant's own, the story of

---

1   Eiseley

a particular place, and a story of the Earth all at the same time. Plants are expressions of networks of integral parts. Each species tells its story in its own unique way, through all of its physical characteristics, growth processes, relationships, and strategies—that is, through its gesture.

Through the plant's flower its story unfolds eloquently, through colors, fragrances, textures, and shapes. The cupped Sagebrush Mariposa, with its crystal-like six part structures, speaks of water in the arid shrub-steppe region. The intricately imaginative Calypso Orchid reflects the complex web of relationships of the deep forest.

Modern life can suspend us in an insularity where the living systems around us seem to have little or no significance. This illusion of separation begins to dissolve when some wild being captures us with its story, its singing. Its medicine attunes us, instructs us in how to survive and flourish in the place where we are, how to mindfully participate within the intelligence of a whole community. Since wild ones thrive while contributing to the whole, nature's medicine helps us resolve challenges in innovative, balanced, and life-enhancing ways. This medicine is the wisdom that helps us to be fully human, belonging to the Earth.

Often I am drawn to a particular flower, and through this encounter I shift into a more coherent, open way of being. Many people have had similar experiences with flowers. There are times when a certain flower (or other aspect of nature) will hold a kind of magic while at other times something else captures our focus. One time in an ancient forest, the white-green shimmer of a Round-leaved Bog Orchid helped to calm, center, and uplift me after an injury. Another time, Heather's pink bells on a mountain pass became a portal through which I could deeply enter the beauty around me, beauty that I had not noticed in my heavy, lethargic state. These and many other experiences of myself and others are similar to those of Edward Bach, the British doctor who discovered the Bach flower essences in the 1920s and 30s.[2] In these years he walked the Welsh countryside, at times in the grip of an uncomfortable mental-emotional state. He would then be drawn to a flower that eased this state, allowing peace to prevail. Other times he focused on a person or type of person whom he wanted to help, and, similarly, he was guided to an appropriate flower.[3]

Some years ago, one of my psychotherapy clients was in as much despair at the end of our session as when he had first arrived. He was a young man who had just experienced yet another devastating loss in a series of losses. Understandably, the world now seemed dark and hostile, ever ready to diminish more of what he loved. As he got up to leave, a photograph on my wall caught his attention. "What are those beautiful flowers?" he asked. Though this picture had hung in the same place for the almost two years that he had been coming to sessions, it was only now that he noticed it.

Aware that his absorption in this picture of a bright meadow of Wyethia was a marked shift from his withdrawn, sullen demeanor during the session, I found myself not only giving him the flower's name but also inviting him to hear its story. I explained that while most other Sunflower types follow the sun in its daily journey across the sky, Wyethia blossoms stay fixed throughout the day and night at the point where the first rays of sunlight touch them each morning. Something in the picture of these sunny flowers had moved him, and he was visibly impacted by their story. As he left I was myself moved, grateful and humbled by his process, by the invisible, unknowable something that had led him to a photograph

---

2   Flower essences are tinctures formulated from fresh blossoms, pure water, and sunlight. Since the 1930s Bach's essences have assisted both people and animals to transform challenging emotional states. His discoveries have continued to inspire new essences and research world-wide.

3   Barnard

of flowers who orient themselves to the first light of day.

The concept of resonance can offer a window of understanding for these examples. Quantum physics reveals that the world around us—including physical objects as well as mental and emotional states—can be understood and experienced as wave patterns, as fluctuations of vibrating energy. Things with the same or similar vibrational patterns resonate with each other; they attract and influence one another. In the above situations, certain flowers could be said to resonate, to correspond in some way, to various intrapsychic states. Even a picture or photograph of a flower can carry its vibrational imprint. Just as a particular piece of music can move some people or affect certain moods, a special flower calls to us at specific times. Resonance is occurring in both cases, for we experience both the music and the flower as patterns of vibrational energy. Our energetic experience of the flower is more subtle than that of sound, of music, so our conscious mind can more easily miss it, but it is no less real or powerful. Attuning to our bodies' sensations and to our thoughts and emotions helps focus awareness on our energetic relationships.

The alchemical tradition recognizes and studies the correspondences of the human and non-human worlds, of inner nature and outer nature. These ancient teachings, cultivated in Egypt and Greece and carried forth in medieval Europe, are often referred to as Hermetics after their legendary transmitter, Hermes Trismegistus. This wisdom points to the universal patterns, or archetypes, that underlie both matter and consciousness. These are the very shapes of existence.

The essence of alchemy is summarized in the dictum, "As above, so below". In other words, celestial patterns are reflected in Earth systems and processes; the microcosm contains the macrocosm; and mind is expressed in matter and matter in mind. Since the tangible corresponds to the intangible, a plant's invisible patterns of energy are accessible to our senses in its gesture. As well as through shifts in our sensations and emotions, we can detect a plant's resonance with us when its gesture, the plant's "story", captures our imagination and infuses life with meaning and perspective.

Carl Gustav Jung (1875-1961), Swiss psychoanalyst and explorer of consciousness, himself a student of Hermetics, discovered that when we are in crisis our psyches generate, or locate, an image, a symbol, which captures us with its numinous quality. Such an image shimmers with divine presence.[4] This image can come in a dream or be something from our waking world. It is a living force that enters us and works as a transformer of consciousness in a very real way. These symbols point the way to wholeness; often they are from the natural world. We can amplify the transformative power of the image by learning all we can about it and experiencing it, by understanding its nature, listening to its story. We can then apply this imaginatively and metaphorically to the challenge before us.

For my client, so caught in despair on that day, perhaps the Wyethia transmitted something about the cyclical nature of darkness and light, about holding faith throughout the darkness in light's eternal return. Perhaps he remembered a way to orient himself towards renewal and rebirth, finding a point of constancy in the midst of sweeping change. The flower's physical attributes and nature corresponded in some way to his intrapsychic state.

When the psyche alights on a flower, recognizing a numinous doorway to transformation, the flower offers a special kind of medicine. Our souls have a natural affinity for flowers. The ancient Greeks pointed towards an intriguing mystery by using the same word, *psyche*, to mean both *soul* and *butterfly*. The soul ever propels us towards transformation. It requires earthy, bodily experiences and immersion in dense emotions as well as sublime purity and inspiration. It prompts us to transmute heavy, painful experiences into spiritual gold, speaking to us through metaphor and symbol, bridging the physical and subtle, embracing both. A butterfly begins as

---

4   Jung

an earthbound creature and makes a transformational leap to the realm of sunlight and air. Similarly, a flower is the plant's metamorphosis, containing the plant's entire journey yet expressing something unprecedented, opening to new relationships, new freedoms and possibilities. The flower is the magnetic counterpart of *psyche*, the soul, the butterfly.

The following narrative portraits of wildflowers of the Cascadia bioregion have arisen from many seasons of following the flowers' magnetic songs. They present observations of each plant's physical characteristics, or "**gesture**", and its web of relationships, its "**habitat and ecology**". Often the descriptions include the plant's herbal, nutritional, and historic gifts. Held imaginatively, metaphorically, and soulfully, this information reveals the flower's "**medicine story**", its essential inner nature, which interfaces with the human journey, offering integration and harmony. These portraits embrace many ways of understanding and relating to plants, and reflect our kinship with the natural world. Flowers speak soulfully and evocatively, commingling the physical and sublime. A flower's fragrance, colors, textures, and designs capture our senses, grounding us in the physical world, in our bodies. They celebrate our senses and the sacredness of our bodies and Earthly life. At the same time their jewel-like colors and compelling scents touch our emotions and imaginations, and uplift and inspire us. The wildflowers, and therefore their portraits, reweave soul and nature, matter and spirit. As we fully open our senses to plants, we begin to perceive their gestures in terms of rhythms, orientations, elements, polarities, movements, hues and tones, metamorphosis, cycles. In other words, we perceive them in terms of the patterns of creation that underlie both our inner experience of moods, thoughts, and emotions as well as outer, physical nature.

German artist, writer, and scientist, Johann Wolfgang von Goethe (1749-1832), developed an approach to the natural world that synthesized disciplined empirical observation with internal perception and response. His work carried the wisdom of the Hermetic teachings, which flowed through Europe as an underground stream during this time of the "Age of Reason". Goethe held profound respect for nature's mystery. He approached nature with the Hermetic understanding of the unity behind her diversity of expression. His work was rooted in tangible, direct experience. After rigorous empirical study of a plant through each part of its growth cycle, he drew his perceptions inward with focused meditation, incorporating imagination and intuition. Through this undertaking he recognized the way the external form of a plant relates to its essential, or interior nature. Through his sensitive and soulful science, plants revealed themselves as dynamic processes participating in larger ones, as intricate relationships to be approached with reverence and patience. Rather than viewing a flower as a static and separate phenomenon, Goethe's approach opens our perceptions so that we begin to experience the flower as a gesture in a larger dance, a gesture that is transforming even as we perceive it.

Goethe's science observes themes that repeat in various ways in each integral part of a plant. As these themes are noticed and reflected upon, they point to the plant's essence. For example, each aspect of a Bleeding Heart plant—including the exquisite softness of its unfolding leaves, its delicate leaf patterns, the color, shape, and enfolded interior of its flower, and its herbal compounds that relieve deep nerve pain—expresses depth and sensitivity, qualities we can apprehend with our senses as well as resonate with emotionally. Through this kind of observation, we can become aware of how each piece contains the whole, and we recognize in the plant aspects of our human experience.

Likewise, each plant is an integral part of its habitat. Each plant expresses an aspect of the overlighting gesture of this habitat. Simpson's Hedgehog and the Large-leaved Lupine each in their own way evoke the lithosol (extremely dry rock-soil) hills where they grow. Their gestures reflect the reciprocal sculpting of water and hills. This landscape speaks of the nature of water, of life's response to water's scarcity. We have returned to recognizing plants as carriers of the songs, the stories, the *medicine* of their places. We become aware of the lands around us as made up of integral places, unique in their expressions, contributing to and interplaying with larger systems. Each place partakes of the guiding wisdom of the universe. We can honor a place and its medicine through awareness of patterns and archetypes, through understanding it as whole, noticing its wholeness reflected in all that arises there. We honor a place through apprehending its soul.

We have inherited a world of isolated pieces in which humans are disconnected from nature, mind from body, spirit from matter. These artificial divisions are evident in our landscapes, for they have been dissected and carved up by our land management policies. When we approach nature only through objective analysis and dispassionate observation, we remove humans from the web of life. Other ways of knowing are short-circuited and we continue the dissection and destruction of nature, including ourselves.

Perhaps we cheer on the demise of humans as the only hope for "saving nature". However, we are inextricably woven into the fabric of the universe. Science illuminates this connectedness when it is valued as one of many windows of perception. Botanical descriptions sharpen our powers of observation and reveal intricate and ingenious adaptations. When we approach botany whole-heartedly, allowing it to enliven our senses, imagination, and intuition, we recognize the same patterns of creation in wild nature, in our bodies, and in our psyches.

Perceiving through the heart, we apprehend unity. We have our own gift, our own medicine, which is itself an indispensable life-giving contribution. Humans have the ability to be completely awe-struck by nature's beauty, to be filled with wonder. We are moved to gratitude, to praise, bowing in reverence.

We yearn to offer back to wild nature something of beauty. Artistic expression allows us to respond sensuously, with the materials of the Earth, with our bodies. Following nature, art reweaves the world, connecting the interior and exterior, physical and spiritual, self and other. In this spirit, artist Karen Lohmann has created the wildflower cards. With chalk pastels these portraits respond to the light, color, textures, shapes, and designs of the flowers. Echoing each flower's essential nature, the cards are a sensual, interactive experience. They draw us into intimate connection with the wildflowers, and with ourselves. Karen's intention is to elicit a deep response that continues to unfold as we relate to flowers, to the natural world.

This collaboration of visual art and narrative honors and celebrates wildflowers and our relationship to them. Specifically, it presents flowers of Cascadia, also known as the Pacific Northwest. These flowers carry the medicines of the many and varied ecosystems contained in this bioregion. Among the ecosystems represented by these flowers are rainforest, shrub-steppe, riparian, prairie, and subalpine. Their lands are arid to extremely wet; they span elevations from sea level to high montane; and they experience moderated marine climates to extremes of temperature. This variety of conditions and geographic features characterizes Cascadia, and it creates a succession of blossoming plants for more than half the year in the State of Washington alone. Many diverse ecosystems lie within Washington's borders; in the contiguous United States, only California contains more.

The thirty-three wildflowers of *Medicine of Place* invite us to interact with the rich and varied ecosystems of Cascadia. They encourage residents and visitors to notice and experience the intricate web of life that supports and sustains us, to receive the medicine of place. The flowers featured in this work all grow in the Cascadia bioregion; more specifically, they can all be found within Washington State. A few are endemic only to a certain ecosystem within Washington. Some grow in other bioregions as well as Cascadia.

Even as we honor these flowers and commend their incomparable bioregion, they and the conditions of their ecosystems are changing. Development and resource extraction continue to destroy ever more habitat. Flowers have responded to recent climate change by blooming earlier and for a shorter amount of time. We have observed a significant increase in the number of mutant flowers beginning in the 2005 season, especially discovering many flowers of five-petaled genera with six petals. With climate change, conifers are encroaching where northern latitudes once kept tree lines low and conspired with abundant snow pack to create vibrant mountain meadows. Meadows such as those that host the world-renowned floral displays of Mt. Rainier are now filling in with young trees. On the other hand, forest fires have increased, leaving in their wake brilliant waves of Fireweed and Arnica. When we rush to label these changes as good or bad, perhaps we preempt our ability to listen deeply and align ourselves with nature's wisdom.

Through the synergy of visual art and written words, Karen and I join with many who advocate for those who speak with other than human voices. We hold the intention for all beings to flourish on our precious home planet. We can be most effective, most of service, through the paths that delight us. What we love instructs us in how to serve it. Service to flowers must be sensuous and soulful, a path of joy and continuous opening, innovative and interdependent, deeply ecological, an offering of light. May the wildflower cards and their stories recall a wider, deeper experience of the world, awakening us to the sensuous, life-giving matrix that holds and nourishes us. Within this tapestry may we discover ourselves, undeniably interwoven.

Chapter Two: Suggestions for Using this Book and Wildflower Deck

> Universal perceptions arise when man strives to understand the powers that work in the silent humility of all Nature.
> —*Paramahansa Yogananda*

A core purpose of this book and card deck is to initiate relationships with native plants, relationships that will develop and deepen over time like cherished friendships. The pastel portraits and stories of the featured wildflowers are enticements to visit and experience them. These thirty-three plants become a doorway to the vast world of flora of the Pacific Northwest, or Cascadia bioregion. At the same time, they lead into a deeper awareness of our selves.

Meeting the Plants

Concise descriptions of each plant's key botanical features and range accompany each narrative. These descriptions, supplemented with a wildflower guidebook, will help you locate each plant and be able to recognize it. Several guides for Pacific Northwest flora are listed in the **References** section at the end of this book.

Timing Visits

Each plant unfolds within a larger cycle, which includes the other vegetation, climate patterns, and geographic features of its habitat. Plant development varies not only according to species; it is affected by yearly temperature and snow pack fluctuations. Such factors as elevation, coastal proximity, and exposure cause plants of the same species to bloom at different times. For assistance with when to visit the plants, the "**habitat and ecology**" section points out some of the other processes that coincide with each plant's blossoming cycle. This perspective is the first step in gaining proficiency in the fine art of timing visits to plants.

Entering a Relationship

As we return to the plants through the seasons, getting to know them in all their phases, we deepen our response to them and discover our kinship with the natural world. The thirty-three wildflowers are portrayed through the language of connection—through art, metaphor, story. Most activities that acquaint us to plants stress identification and classification. While these are useful tools and can help develop observation skills, they tend to eclipse other ways of knowing and reinforce a fragmented world view. Reaching for a plant's name can be a way of capturing and collecting it, effectively over-riding the sensations and inner experiences that arise as we first encounter it. Allowing our responses, however subtle or perplexing, to unfold can open new perception.

Naturalist and educator Paul Krafel writes that when he encounters a wild animal, he has shifted from asking, "what is its name?", to "what is this animal doing?" The second question, rather than referring us to an external authority, calls upon our own powers of observation. It brings us into the present moment and leads to more questions, opening into an ever-unfolding process of discovery.[1] The "**habitat and ecology**"

---

1   Krafel

and "**gesture**" sections of the wildflower portraits can be understood as a response to the question, "what is this plant doing?" Surprisingly, this question keeps expanding awareness. Each plant is growing in certain directions, displaying particular designs, engaging with other beings, responding in its own ways to water and sunlight.

Empirical observations lead to discovery of themes and patterns, which point to the essential nature of each plant; each aspect of its gesture expresses the plant's essence. Following Goethe's approach, we discover universal patterns that apply to both inner and outer nature. The "**medicine story**" of each wildflower unfolds these patterns, discovering the plant's function within its form, recognizing its essence reflected in our human stories.

The Flower Essence Society of Nevada City, California draws on Goethe's work in their plant attunement processes, through which they bring into focus the essential soul medicine of each flower.[2] **Table 1**, on page 17, outlines suggestions for encountering and opening awareness to plants. These suggestions incorporate aspects of the Flower Essence Society's plant study process, and offer guidance in developing a grounded and soulful kinship with the wildflowers.

Self-Reflection and Soul Work

In the practice of being with a plant, we deepen into communion, both with the plant and with ourselves. The plants arise from the same archetypes that shape our interior nature. We notice which plants call to us, or resonate with us, and notice the ways they affect us.

This process can be explored using the wildflower cards. As artistic portraits, the cards carry an aspect of the vibrational patterns of the living plants. They are convenient tools for insight, awareness, and divination. The wildflower cards lend themselves to creative play, and invite discovery of many applications.

---

2   Kaminski, Katz

One can chose cards randomly or on the basis of affinity as part of a daily meditative practice. The cards can be called upon for troubling emotions or problems, or to gain insight. When drawing cards at random, or "blindly", the unconscious wisdom of the body guides us to what resonates with our present state. In choosing the images that most attract us, we are responding more consciously to this resonance, following our intuition, much as we do when we are out in nature.

The next step is to sit reflectively with each selected flower portrait, receiving it visually and imaginatively, allowing inner experiences to surface. Reading the flower's narrative, becoming acquainted with its physical expressions, habitat, and relationships, as well as its "**medicine story**", can expand awareness and spark insights. Journaling and discussion can further help you to receive the flower's medicine.

The wildflower cards can be used in the same way as Tarot or other divination cards. They can be drawn and arranged in various spreads, or formats. A simple and powerful three-card layout is illustrated in **Table 2** (page 19).

Mindful encounters with the plants in nature and processes with the cards affirm the correspondence between our intrinsic nature and wild nature. Wilderness mystic and advocate John Muir recognized this interface of the psyche with the sensory world around us. "I only went out for a walk," he wrote, "and finally concluded to stay out till sundown, for going out, I found, was really going in."

Pathways of Connection

The *Medicine of Place* artistic and written portraits are designed to open both physical and conceptual doors. They point our steps to encounters with plants and open awareness to each plant as a web of connections, in which we participate. These materials lead beyond "learning about" nature to experiencing ourselves within nature, connecting the inner and outer worlds. The "environment" is no longer just a world "out there"—outside the door and outside of us—which we must in some way control.

Our responses to nature are often not noticed let alone acknowledged, valued, and examined. At this point in history, we are striving to awaken what is already inside us. We are remembering the ways of knowing of our child selves, of our ancestors, and of humans still connected to the rhythms of the Earth. Overall, this book and the wildflower cards are encouragement to follow intuition, amplify spontaneity and wonder, and surrender to the magnetism of the flowers. In this way we open our hearts to the revelations of the plants.

Table 1

Suggestions for Intuitive Connection with Plants

FOCUS

1. Take a few moments in silence to ground into the Earth, and to fully inhabit your body.
2. Open and activate the heart through breathing into your heart center. Throughout the process perceive through the heart, refocusing awareness with the breath.
3. Allow a question, challenge, or emotion to emerge. Or, set an intention to best serve your general well-being or to be of service to another.
4. Express gratitude to the plants and to the natural world.

ENCOUNTER / COMMUNION

1. Walk slowly among the plants, noticing what captures your attention. Does a particular plant seem to be calling you?
2. Sit with this plant allowing spontaneous enjoyment and curiosity to guide you.
3. Open all of your senses, taking in as much sensory information from the plant as possible. You may wish to record these empirical observations in a notebook. Sketching the plant can sharpen awareness. Which senses are most activated by this plant?
4. Take your time, staying open to whatever unfolds, honoring the organic flow of relationship. Feel the exchange of energy between you and the plant through your heart.

REFLECTION

1. What was the first or primary quality that captured your attention and drew you to this plant?
2. Does this quality hold a particular meaning for you? Describe this quality using many adjectives.
3. How did your encounter with this plant impact your physical state of being. Scan your body for sensations, noticing their quality and location.
4. Similarly, check in with your emotional and mental states. Were any thoughts or emotions stirred up or calmed?
5. What about this plant seems to be akin to you? What seems very different, perhaps alien?
6. Reflect on this plant's characteristics in terms of polarities, elements, and archetypes.
7. Does the plant evoke images, memories, or stories? Imagine this plant in your heart center and notice what emerges.
8. Express your impressions of your plant encounter artistically with a movement or gesture, visual art, music, or writing. Is there a message?
9. Revisit your original question or state of being. Describe any shifts or insights.
10. Through which channels did information most readily come to you?

CLOSURE

1. Express gratitude to the plant and your surroundings.
2. Take a few moments in silence to integrate your experience.

Table 2

Three-Card Layout

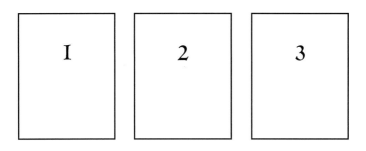

Card 1: The problem or issue. Provides a way of framing your challenge, question, or event in order to gain clarity, understanding, and perspective.

Card 2: Action. Guides you in discerning the appropriate response to the situation. Points to what is required of you.

Card 3: Spiritual insight. Reveals the higher wisdom available through this situation. Helps you grasp the evolutionary edge you are facing with this challenge, and brings into focus the soul lesson that wants to emerge.

Chapter Three:
The Medicine of Flowers

> The weight of a petal has changed the face of the world and made it ours.
>
> *—Loren Eiseley*

> If we could see the miracle of a single flower clearly, our whole life would change.
>
> *—Buddha*

The plants that produce seeds within the vessel of a flower are the most recent arrivals of the plant kingdom. The adaptation of the flower has caused a great leap in plant evolution. Because of their efficient and ingenious reproductive strategies, flowering plants—or angiosperms—appear to have an almost unlimited potential to evolve and flourish. Once the first flowers appeared—about one hundred million years ago—they spread with unprecedented speed and efficiency across the Earth.[1]

Similarly, the first humans precipitated a rapid expansion of previously unknown proportions, soon inhabiting the far reaches of the planet.[2] Humans and flowering plants are both expressions of transformational

---

1 Eiseley
2 Swimme

power; both represent an almost inexhaustible creativity in response to life challenges. These plants are like our twin soul, manifesting the same archetype, yet in a very different way. Unbound by instinct, humans engage in unbridled experimentation, continually developing new tools and technologies. Hopefully we learn through trial and error. On the other hand, the innovation of the flower, its ingenuity and achievements, are contained within an interdependent matrix. Each gesture of the wild native plants sustains the integrity and harmony of their communities. The flowering plants reflect back to us the power of creative actualization, while at the same time, they instruct us in the wisdom of balance, interdependence, harmony, and integrity.

The essence, or medicine, of each flower species is a variation on the collective gifts and archetype of the angiosperms. Each native species expresses an eloquent resolution to its particular environmental and growth challenges while contributing to the balance and harmony of its habitat. For example, Fireweed's highly effective seed dispersal strategies make it a prolific pioneer in damaged areas. It quickly lays an underground network of root stalks, reweaving the life-supporting structures of the soil. Fireweed's vigor, rather than overwhelming and excluding other life forms, is an invitation to life's renewal after devastation, just as its bright magenta flowers rekindle our spirits.

Taking a closer look at the role flowers play for their plants leads to a deeper understanding of their collective gifts. Flowers are highly magnetic. Irresistible to many living creatures, life vibrates and hums around them. They attract life-giving processes that resolve challenges for the plant and simultaneously enhance the ecosystem. The flower opens the plant to new possibilities for communion with the surrounding world. Its designs, colors, fragrances, and nectars draw many types of insects, and these in turn nourish the soil and other life forms. Animals and birds feed on the flower's pollen, nectar, seeds, and fruit. Flower medicine reveals to us previously unimagined relationships and possibilities for creative collaboration. Flowers instruct us in attracting what enables us to flourish and contribute to the greater community.

In flowering, the plant makes a transformational leap. As leaves approach the tip of the stem where a flower bud will appear, they simplify and become smaller. It is as though the plant contracts just before flowering, gathering energy for this unprecedented task. The flower's opening is an explosion of creativity, unforeseen in the plant's previous growth. As a plant comes into flower it reflects our own developmental processes and transformations, which include withdrawal as well as expansion.

Through flowering, the plant shape shifts. Flowers reflect the qualities of the elementals, of pollinators and other lifeforms. The shape of a blossom and the shape of a bee inform each other, and the sun, wind, and water are reflected in floral gestures. Flowers live at the edge where separate life forms become a communal dance. They evoke our mystical hearts.

The collective archetype of flowering plants comes to life through the kaleidoscope of thousands of species. Each species in its own way moves and inspires us. By exploring each one's gesture, ecology, and adaptive strategies, we begin to integrate our responses to a particular species with deeper understanding. The following sections look at the ways color, sacred geometry and form, ecology, and the plant's connection to humans reveal a species' essential medicine. These qualities are the language through which a plant tells its story. Through its sensuous, tangible forms, a plant reflects its inner nature; its folk stories and application to human needs are all aspects of its archetypal footprint.

Color

Floral colors, whether soft or vibrant, are light-filled, alluring. Color is a quintessential gift of the flowering plants, for their development expanded and brightened the palette of the Earth. Preceding flowers, the earliest plants were dependent on water for reproduction. They extended only ribbons of green along shorelines. Eventually, conifers contributed areas of rich shadowy green. In contrast, the flowering plants unfolded a vibrant, multicolored mantle across the Earth. Leafy bowers and grasses spread fresh modulations of green, and the flowers added the brilliant hues that previously had only appeared in the sky. The rainbow colored flowers have joined the Earth and heavens.

The impact of color on our health and well-being is an ancient science. Each color has its own frequency and wave length. Moving along the visible spectrum from red to violet, colors become higher in frequency and shorter in wave length. Through its vibrational quality, each color impacts us in a different way. Colors affect our endocrine glands and therefore the secretion of hormones. This helps to explain their influence on mood, appetite, and healing capacity. Optometrist Jacob Liberman explains that the prevalent colors of each seasonal landscape prompt corresponding hormone secretions in animals and humans. These hormones stimulate seasonally appropriate behavior and metabolic activities.[3]

Of the visible colors, red has the slowest frequency and longest wave length. It is the low edge of the spectrum, the emergence into the visible. Predominantly red flowers can stir body sensations, energy, and intensity. Red is physically stimulating and can catalyze processes and actions. Consider the invigorating effect of a meadow filled with vivid red Paintbrush.

On the other end of the spectrum, violet is the most rapidly vibrating color with the shortest wavelength. It has a very sublime quality and can uplift and inspire. Violet occupies the upper edge of visible light. In alchemy the violet flame has the power to transmute substances or states of consciousness to a higher level. Violet is the leaping off point for transformation. Many of the plants that first re-enter a destroyed area, such as Fireweed and Foxglove, have violet flowers.

Green is the medium through which sunlight becomes nourishment,

---

3   Liberman

making possible the proliferation of life on Earth. Green flowers exalt the life-giving mystery of green plants. In the middle of the visible color spectrum, green blends warm and cool colors. It is harmony and balance. The heart chakra, the middle chakra where Earthly and cosmic forces merge, is associated with green. Green plants have a special affinity with the heart, which carries the wisdom of interconnection, the knowledge that each life is valuable to the whole. Green flowers, such as Green Bog Orchid and Mountain Sweet Cicely, offer a soothing, balancing balm that can heal physical and emotional disturbances. Green harmonizes while preserving the integrity of diverse qualities, and restores coherence throughout the system.

Both yellow and white flowers shine with abundant light. Yellow flowers are the brightest mirrors of the sun. They bestow radiance to their surroundings, and inspire us to express brilliance. White is the reflection of all colors. White flowers are often used in sacred ceremonies; they emanate holiness, the one light behind the diversity of creation.

In relating to flowers, be aware of the effects of various colors on different areas of the body, on physical sensations, breath, moods, and thoughts. Consider the correspondence of the chakras, from root to crown, with the ascending colors of the rainbow.

Sacred Geometry and Form

Since the structures of the natural world also shape our inner world, the patterns of manifestation, energy, and awareness are the same, and they reveal the unity of Creation. The tangible expression of these patterns is known as sacred geometry. The archetypes that are the structures of existence are ineffable, but their meanings are encoded within sacred geometric shapes. Humans have woven these patterns into ancient stone circles, great cathedrals, and other places of power and worship. We respond strongly to flowers because they display universal sacred forms. Because of the correspondence of inner and outer, these geometric patterns resonate with levels of awareness.

Sacred symbols which appear in flowers include: the Star of Bethlehem, or six-pointed hexagram star; the five-pointed pentagram star; and the Celtic Cross. These symbols help identify a plant, since traditional botanical classification is based on the observable patterns of its flower. Classification, including family and genus, reflects geometric, and therefore archetypal, patterns.

The flowering plants are broadly divided into plants with one seed leaf, called monocotyledones (monocots), and plants with two seed leaves, eudicotyledones (eudicots). Monocot flowers have parts in threes and sixes. The monocot families Liliaceae and Iridaceae have flowers with three petals and three sepals. This forms a hexagram, or Star of Bethlehem, a figure composed of two interlocking triangles. (See **Table 3** on page 29.) The Star of Bethlehem represents the interpenetration of heaven and Earth. It points to the interweaving of the subtle and the tangible, the divine and the material. Matter is sacred, and spiritual principles seek incarnation.

The theme of connection between dimensions, symbolized by the upper and lower realms, is repeated throughout the gestures of the monocots. Plants of the Lily and Iris families tend to grow rapidly upward on vertical stalks. Lilies, Irises, and Orchids are minimally rooted; many have bulbs and tubers that can survive when removed from the ground and stored. Though their life force is cradled within the Earth, it is also transcendent. In Greek mythology, Iris is the messenger of the goddess Hera. Often appearing as the rainbow, she bridges the worlds of deities and mortals.

The symbolism of a conduit continues in the monocots' correspondence to water, which also displays a hexagonal structure in its crystalline forms. Water connects Earth and sky through its hydrological cycle. In its solid, liquid, and gaseous forms, it spans dimensions. Water carries the archetype of the divine messenger in that it mediates between the nonphysical realm and living systems. Its molecules, because of their unique structure and properties, receive the imprints of vibrational patterns.[4]

---

4   Hall

Water transmits these patterns to living systems, which decode them, gleaning instructions for vital processes.[5]

The monocots, with their upright stems, mirror the vertical aspect of the human gesture. Our bodies are conduits between sky and Earth. Three and sixfold flowers point to our existence beyond the physical, and at the same time to the sacredness of physical manifestation. They reflect our potential to anchor spiritual principles into the material world, consecrating Earthly life.

The flowering plants with two seed leaves, the eudicots, frequently display fivefold patterns in their gestures. Rosaceae flowers, for example, have flowers with five petals, inscribing a pentagram. (See **Table 3**.) This five-pointed star is another sacred form, encoding the mysteries of growth and regeneration. From within a pentagram a golden spiral can be uncoiled. This spiral is the movement of life's unfolding processes.[6] The

---

5   Greene

6   See Schneider. Each point of intersection of the lines of a pentagram divides the line into two parts which are related to each other through the golden mean. The golden mean proportion (also known as phi, or 1.618...) is the ratio in which the relationship of the parts to each other is the same as the relationship of the parts to the whole. This is a ratio of beauty and harmony through which nature generates self similar growth, ever creating new forms, while each maintains a relationship and correspondence to each other and to the whole. The golden mean appears everywhere in nature's architecture, from the bodies of humans and insects to the structure of the solar system.

Because of the pentagram's golden proportion, we can continuously generate new five-pointed stars from within the centers of each successive one, or we can outwardly expand new ones from any succession of star points. This is an infinitely generative process, because consistent proportion—and therefore balance—is maintained as growth continues to emerge from within the system.

Connecting the same point of any succession of five pointed stars inscribes a golden spiral, the spiral of life that grows from within through self-accumulation.

golden spiral generates from a still central point at infinity and it maintains the same proportion (known as the golden ratio) as it grows. The process of a plant unfolding, opening from seed to shoot to leaf to flower to seed, inscribes a golden spiral. The new seed opens and the process begins anew, life ever emerging from the infinite emptiness within.[7]

Fivefold flowers, such as those in the Rose family, express the inexhaustible self-generating power of the Earth. They connect us to the beauty and harmony of nature's processes. Rosaceae and many other families with five in their gestures are firmly rooted, displaying robust growth and propagation characteristics. Their gesture is more horizontal, spreading over the land, as compared to the vertical tendency of the monocots. Fivefold plants prompt us to fully inhabit the Earth and our bodies. Through attuning to the lovely five-petaled flowers we discover their correspondence to our own bodies, whose proportions are the golden ratio. In recognizing the beauty and miracle of our bodies we are encouraged to make use of our five appendages, our fivefold feet and hands, and our five senses to take action, create, engage in Earthly life, and to honor our connection to the beauty and regenerative power of the Earth.[8]

Flowers from both the Evening Primrose and the Mustard families have four petals, forming a Celtic Cross. The number four reflects the four directions through which we orient ourselves to Mother Earth. Adding four more directions in between these cardinal points yields eight (Evening Primrose flowers have eight stamens), which symbolizes infinity and represents the infinite points on the sacred circle of the Earth. Each being has a unique perspective according to its place on the circle of life. We and all of creation are formed of the four elements of the Earth

---

This spiral maintains the golden mean proportion, and expands infinitely from a still central point that is approached but never reached. Golden spirals can be recognized in animal horns, shells, storm patterns, galaxies, and plant growth.

7   Adams, Whicher
8   Schneider

Mother: earth, air, fire, and water. Four is the number of the mother substance and the holy matrix of Creation in which we are held.[9]

Along with color and geometry, a flower's orientation on its stem, its symmetry, its resemblance to other shapes and life forms, and its overall visual design are expressive and meaningful. Many flowers open heavenward, some spread like satellite dishes, others are cup-like. Certain flowers bend toward the Earth like bells or with petals reflexed back. Still others face horizontally on their stalks. Each of these orientations conveys something different: openness, containment, offering, receptivity, protection, restraint, exuberance, dynamism. Is each flower bestowing or absorbing something from the ground or sky, from the four directions?

The face of a flower is often symmetrical, reflecting the perfection of celestial bodies and rhythms. Many Asteraceae flowers resemble mandalas that draw us into their centers. These flowers, which are made up of many florets, express the integration of parts into a focused, coherent whole. Some of them, like Arnica or certain Daisies, are used for herbal and homeopathic preparations to assist with shock and trauma. They help draw in, ground, and focus the life forces. The umbrella-shaped Apiaceae flowers, on the other hand, visually draw us up and out. This centrifugal movement is associated with ecstatic experiences, pain relief, and heightened sensitivity to subtle dimensions.

Flowers such as Mimulus and Penstemon are bilaterally symmetrical, like the shapes of animals. In these gestures we recognize emotions—anger, fear, laughter. The Latin name *Mimulus* refers to this *mimicking* quality of the flowers of this genus. Many flowers enclose an interior space, as animal and human bodies and organs do. We see this in Bleeding Hearts, many Orchids, and in trumpet or bell-shaped flowers. They form a hidden chamber, reflecting the inner sanctity of the self. These flowers express emotional sensitivity, vulnerability, and all that we guard within.

---

9   Ibid.

Table 3

Sacred Geometry and Flowers

Hexagram; 6-pointed Star Star of Bethlehem; Star of David; Seal of Solomon	As Above, So Below Connection of Spirit and Matter Involution and Evolution	Lyall's Mariposa *Calochortus lyallii*
Pentagram; Five-Pointed Star	AB/BC = BC/AC = Phi = 1.618... = Golden Ratio Nature's Harmony, Beauty, and Regenerative Power	Large-leaved Avens *Geum macrophyllum*
Celtic Cross Planet Earth Symbol	Four Cardinal Directions Four Elements	Fireweed *Chamerion angustifolium*

Through flowers appear hearts, shooting stars, slippers, heads of elephants and beaks of birds. Orchids masquerade as exotic moths. This carnival of forms reveals the flower's relationship to pollinators. With its imagery and correspondences the flower points beyond itself. Curiosity and poetic vision uncover connections that may at first seem distant and unrelated.

The elementals are portrayed in floral gestures. The cup of the Mariposa and the droplet of the Bleeding Heart conjure the presence of water. Many blossoms and seeds conspire with the wind, imitating the winged world, weaving airiness into their forms. Asteraceae genera reflect various aspects of the sun's glory. Large, pendulous flowers and woody stalks express earthiness. Hermetic wisdom recognizes the correspondences between the physical elements earth, air, water, and fire and our sensual, intellectual, emotional, and energetic qualities.

Ecology

A plant cannot be separated from its web of relationships and the environment in which it is found. A native plant is a keeper of the knowledge of how to flourish in and enhance that environment, a keeper of the medicine of right relationship to its place. Within the plant's web are other plants, animals, minerals, elementals, seasonal cycles, land formations, climate, pollinators, and predators. Each strand fits into overarching themes. Design qualities relate literally and metaphorically to environmental ones, as well as to emotional and spiritual principles. Each piece is a microcosm of a greater process.

For example, the Sagebrush Mariposa Lily grows from a fleshy bulb, displays hexagonal designs in its cup-like flower, thrives in an arid environment, and blooms after rising temperatures have withered other vegetation. These qualities indicate this plant's ability to contain water and sustain itself though moisture is scarce and heat intense. This Mariposa is an emblem of nourishment and beauty that endure through harsh times.

Plants are instructed by their habitats, shaped by a climate, season, or time of day. High mountain plants know about extremes, about seizing brief windows of opportunity; lowland plants extend deep roots and grow luxuriant during their longer seasons. Prairie and meadow plants have a different relationship to light than those that grow in the forest. The Grass Widow that blooms when winter still reigns has a different wisdom from the Explorer's Gentian which shines at the edge of Autumn. Some flowers, like Simpson's Hedgehog, protect their pollen by closing at nightfall or when clouds threaten rain. On the other hand, certain species of Evening Primrose open only from evening to early morning or on cloudy days.

As we explore the flower in relationship, we see it as part of a field of possibility in which many things arise together, co-creating one another within a larger story.

Connection with Humans

A plant's archetype is also expressed through its interface with the human world. Flowering plants have captured the human imagination and been of practical and spiritual assistance throughout history. We have an intuitive connection to a plant's archetype, which shapes our interactions with the plant.

Plant names reflect the ways a flower makes itself known to us. A plant's common names often create a kind of folk story, a collective imagery. Even botanical names can arise from an intuitive sense of the essential nature of a plant. Names can emphasize certain qualities as well as lead to further questions: "Swamp Lantern" indicates luminosity in dark places; "Salmonberry" connects this plant's fruit with a fish that undertakes a heroic journey of homecoming and sacrifice, bringing ocean nutrients to terrestrial forests; "Calypso Orchid" relates a flower to a Greek sea goddess-nymph.

Many plants are featured in stories and folklore that have been handed down from older, more Earth-based cultures. These are valuable windows into plant medicine since they come through the interconnection of people, land, community, and the imagination. Many stories about Pacific Northwest plants belong to Northwest Native people.

Both historic and contemporary use of a plant for healing and for tools and artifacts illuminates its essential nature. Often the physical medicine and emotional healing offered by a plant express different octaves of the same quality. For example, Bleeding Heart has been used as an herb for relief of deep nerve pain and as a flower essence to ease the emotional pain of grief. With its textures, forms, and colors this plant expresses depth and sensitivity. Herbal tinctures of Oregon Grape root are used to enhance the immune system, while the flower essence promotes healthy boundaries that allow for intimacy. The honey-scented flowers surrounded by prickly-edged leaves dance between sweet offerings and self-protection.

These many facets of flowering plants are explored within the following thirty-three monograms of Cascadia wildflowers. As we expand perception we journey further into the beautiful mystery of the flower. Its wonders begin to reveal the magical order of creation. The face of a flower reflects the amazing world inside of us. The flowers are a testimony to the beauty of life, and the plants can help connect us to this beauty even when all seems lost. Paramahansa Yogananda writes: "Every saint who has penetrated to the core of reality has testified that a divine universal plan exists, and that it is beautiful and full of joy." The flowering plants express this divine plan; they reveal the universal heart. Perhaps each flower is an emissary whose job is to bestow one unmistakable experience of beauty.

Part II: The Wildflowers of Cascadia: Thirty-three Monograms

He who sees into the secret inner life of the plant, into the stirring of its powers, and observes how the flower gradually unfolds itself, sees the matter with quite different eyes—he knows what he sees.

—Johann Wolfgang von Goethe

Behold the herbs! Their virtues are invisible and yet they can be detected.

—Paracelsus

Knowledge this man prizes best / Seems fantastic to the rest: / Pondering shadows, colors, clouds, / Grass-buds and caterpillar-shrouds, / Boughs on which the wild bees settle, / Tints that spot the violet's petal, / Why nature loves the number five, / And why on the star-form she does thrive. / Lover of all things alive.

—Ralph Waldo Emerson (from "Naturalist")

Grass Widow

Olsynium douglasii

IRIDACEAE

Hope begins in the dark, the stubborn hope that if you just show up and try to do the right thing, the dawn will come. You wait and watch and work: you don't give up.

—*Anne Lamott*

Botanical Description

A deciduous plant with magenta flowers and grass-like foliage, growing 6-10" tall. One to 3 flowers, 1 ½ inch across, are borne on pedicels. Flowers have 6 similar tepals. The three stamens are shorter than the elongated style, with its 3-pronged stigma. Ovary inferior.

Range

In Washington: Most readily found east of the Cascades; grows in the middle to east Columbia Gorge. Also found in the northern rainshadow areas of Puget Sound. **Other Areas:** Abundant in southeastern British Columbia, including Vancouver Island. Extends south through eastern Oregon to California and east to Idaho and northern Utah. Associated with low to middle elevations, but the *inflatum* variety, which tends to grow in the easternmost parts of the species' range, can be found up to 6,000 feet.

Habitat and Ecology

At the very edge of winter the Grass Widow begins blossoming, rousing dormant land with satiny color. Also known as Satin Flower or Purple-eyed Grass, it has been sighted even in January; by late February and March it is abundant. Northwest flower seekers enjoy an extended season thanks to the precocious Grass Widow and micro climate areas of the Columbia Gorge such as the Catherine Creek drainage. In the Catherine Creek area, eastern Washington sunlight and temperature-moderating moisture from the Pacific Ocean conspire to dissipate winter's chill.

Grass Widows grow in open places such as prairies and rocky slopes, in habitats with sagebrush, juniper, pine, or oak. Their essential requirement is for abundant early season wetness that will completely drain as spring progresses into summer. These plants thrive in a thin layer of soil covering rock or gravel, through which water can percolate. Summer temperatures that exceed 100 degrees and perennial winds enhance the moisture-leaching capacity of Grass Widow's preferred environments. Yet, to encounter this plant during its bloom time is to appreciate its affinity for water. Where water pools in the contours of the land, clumps of Grass Widows congregate. In spring heat, flowers and foliage wither, and the plant draws its vital force back into its fleshy rhizome.

Gesture

Grass Widow's blossom opens skyward then bends on its pedicel to face the horizon. Thus the slanted sunlight of late winter / early spring perfectly illuminates its floral face. Its six symmetrical tepals, vivid violet to magenta, create a bright mandala with its three golden anthers.

Grass Widows always grow in clumps and spread quickly from rhizomes, splashing the winter-brown meadows with the colors of resurrection. Illuminated blossoms create pools of violet beneath bare-branched oaks. The Grass Widow bestows a celestial lightness when winter's dense, contracted states prevail. Rapid vertical growth and flowers inscribing a star of Bethlehem suggest its role as a conduit between the upper and lower realms. (See Chapter 3.) Suffusing Earth with sky colors, this plant is an emissary from a brighter season.

Grass Widow's leaves and stems are thin and sharp-pointed, a leaf blade rising above each flower, and each tepal bears a pointed tip. Plants often blend with meadow grasses, finely structured, attuned to the wind. Hanging from their pedicels, the blossoms quiver in the gentlest breeze. Bell-like, they ring violet waves into the sleeping land.

Medicine Story

The Grass Widow first arrives in the dark time, heralding the coming of light and color. It bears the promise of emergence from winter. Light and airy in essence, it appears in a watery world and dies back to sustain itself through a fiery one. The flower's red violet color uplifts, arousing joy and inspiration. Violet appears in certain threshold events, such as rainbows and the sky at dawn and dusk. It is the edge of the visible spectrum, the highest frequency color, the stepping off point. In alchemical traditions, the violet flame is associated with transmutation, with the ability to purify and quicken dense material into more refined and light-filled states. Other violet to magenta flowers, such as Fireweed and Foxglove, inhabit transitional land that is re-emerging from destruction. They are floral alchemists like the Grass Widow, who transmutes winter bleakness into vibrancy.

The seasons reflect the continual cycles, small and large, that move through our personal and collective lives. Just prior to emergence from a dark and heavy period, our ordeal seems intractable and unending, more unbearable than ever. The medicine of Grass Widow can be of assistance. Make a pilgrimage to the blossoming plants; invoke them through imagination, story, or pictures; take the flower essence; or bring the color violet into your surroundings. This will quicken your energy field. Ask what you need to burn away in order to break free to another level. This question can apply to a small interaction in the course of a day or a big life cycle transformation.

Many cultural traditions consider our present time to be one in which many cycles at once are drawing to a close, a time of great transition. This threshold challenges us to collectively evolve, transmuting what no longer serves, yet despair often threatens to overwhelm. The medicine of Grass Widow can be found in those visionaries who hold the bigger picture, who are emissaries of a brighter age. We can each tap into this medicine, reaching forward to borrow levity and inspiration from what has not quite begun. In so doing we help to birth a new light-filled time.

Salmonberry

Rubus spectabilis

ROSACEAE

You cannot pick out anything without discovering that it is hitched to everything else in the universe.

— *John Muir*

Botanical Description

3-12 ft. high shrubs growing in dense thickets. Leaves: alternate, compound, 1-3 inches long, double-toothed, pointed at tip and round at base. Leaves are green above, covered with dense, whitish hairs beneath. Flowers: 1-2 inches wide, single. 5 dark pink petals, 5 green sepals, numerous stamens. Fruit: blackberry-shape and size; yellow, orange, or salmon-red.

Range

In Washington: Grows up to 5,000 ft. from the west slopes of the Cascades to the Pacific coast; common in the Puget Sound area. **Other areas:** Native to Alaska, extends south through British Columbia to the redwood country of northwest California; eastward to Idaho and Montana. Especially common in Oregon's Willamette trough.

Habitat and Ecology

Salmonberry requires water-saturated ground and bright splashes of light, if not full sun. In woodlands it forms thickets in wet gullies and bog edges, places where sogginess has curtailed tree growth. Swamp Lanterns, Horsetail, and Devil's Club often grow nearby. Stream sides and riverbanks offer opportune conditions for dense Salmonberry thickets. Its shrubby plants colonize logging sites and burns, marshy meadows, wet mountain slopes, and avalanche tracks.

Salmonberry thickets provide shelter and forage for many wild creatures including foxes, rabbits, Douglas squirrels, sparrows and other ground-loving birds. Their shade helps waterways maintain the cool temperatures integral to the health of salmon and other indigenous aquatic life. Salmonberry flowers as early as late February in the Puget Sound lowlands and its berries are among the first to ripen, ushering in the season of abundant food sources. The lyrical song of Swainson's Thrush is first heard as the berries begin to ripen, and in many Northwest native languages the name for this thrush translates to "Salmonberry Bird".[1]

The Osoberry blossoms may be the first to appear among the lowland shrubs and forest flowers, but the striking Salmonberry is soon to follow, with Trilliums, Swamp Lanterns, and Red-flowering Currant lending their bright notes to this early spring chorus.

Gesture

True to its species name, *spectabilis*, the Salmonberry is visually stunning in every season. In winter its coppery canes and twigs are easily recognized as they cast a satin glow amongst other branches. The canes have a scattering of prickles and shaggy bark, while the upper, newer growth is smooth and shiny. The twigs zigzag at every leaf node; dormant Salmonberry thickets are alive with movement, like a curving stream, rushing water, darting fish.

As a member of the Rose family, the blossoms have five separated petals surrounding a spray of creamy yellow to white stamens. The petal color is a remarkable shade of red-purple, magenta, or deep pink. The color is especially outstanding among a genus of otherwise pale flowers. Spring green sepals strike a vivid contrast to the petals.

Petals and leaves are crinkled like pink and green crepe paper. As blossoms open, thickets sparkle with jeweled droplets of color. Furry bees crawl underneath the downward-hanging "faerie caps", and the flash of the Rufus Hummingbird is not far off. Springtime Salmonberry conjures an enchanted world of faerie-folk, a joyful celebration of life's renewal.

When petals become pink confetti on leaves and trails, they alert hikers to check for ripening berries. The berries introduce a new and equally striking palette of colors. They can be yellow, but are usually bright orange or deep red orange; now they are the jewels, gleaming among rich summer greens.

---

1   Pojar et al.

In color and form the bright orange berries look identical to a cluster of salmon eggs; the deep red ones are the color of salmon flesh. One can imagine how the plant's common name arose from the appearance of the berries. Some associate its name with Salmonberry's use as a food source. The tender stem shoots were eaten with salmon or salmon roe by Northwest native people, the astringency of the shoots complementing the richness of the fish.[2]

Salmonberry has both an irresistible beauty and vigorous growth patterns. It offers food for all the senses, nourishment for the heart and the physical body. The flower's fivefold geometry expresses harmony, beauty, and the infinite regenerative power of nature. (See Chapter 3.) While the fruits and flowers bring light and festivity, the roots are strongly entrenched in the Earth. A network of rhizomes spreads a tangled fortress of plants. Not as formidable as some dagger-thorned members of its genus, still Salmonberry has an unquestionable presence.

From compounds in soil and air, through mysterious alchemy, Salmonberry produces its remarkable colors. The colors of the flowers and berries are distinctive even from each other, and one might wonder what invisible processes connect them.

Medicine Story

The poetic language of common names often juxtaposes separate phenomena, revealing relationship. Such is the case with "Salmonberry". The plant's berries imitate salmon eggs; its shoots provide food that accompanies salmon. Salmonberry finds ideal habitat along streams and rivers, and its connection to water and to salmon is echoed in the zigzag of its twig growth. Salmon require cool water for many reasons, including the abatement of disease organisms. Cool water temperatures create an optimal environment for salmon's nutritional sources and ensure the

---

[2] Mathews

health of eggs. Temperature affects the very structure of water, with coldness enabling better vortex formation. Salmon depend on strong vortexes to be able to swim upstream; their elongated egg-shaped bodies are perfectly engineered to glide through these water forms.[3] Salmonberry's vigorous growth, which quickly produces fortress-like structures, and its attraction to waterways enlist it as a perfect shade-provider for salmon streams. Salmonberry's language of form, color, and gesture unfolds the mysteries of nature's perfectly orchestrated web of relationships.

In their life journeys salmon connect mountains with the sea. They transport nutrients from the rich oceans to inland fresh waters; they themselves become food for birds, mammals, and plants. Marine compounds are woven into the fabric of Northwest forests.[4] Salmonberry, like salmon, connects many living systems with its matrix of support.

Salmonberry invites you to view your life with poetic eyes, with the vision of connection. Its medicine reminds us of the integral nature of our lives. Plants in the Rose family encourage us to fully embrace life, to revere the profound opportunity of being on the Earth. Salmonberry flowers and fruits awaken our senses to Earthly delights, and its anchored roots and vigorous growth demonstrate strong physical presence. Close observation of this plant reveals that no life is isolated or inconsequential.

Attunement to Salmonberry allows us to consider an individual life from a deep ecology perspective. In making a decision we bring to awareness the larger processes that may be set in motion. Salmonberry can assist us when we feel powerless, believing we have little impact. We withdraw from life when we lose sight of our connectedness and the indispensable roles we each play. Salmonberry medicine instructs us to honor the sacred fabric of life through honoring each life thread, including our own.

---

3 Greene
4 Jay, Matsen

Swamp Lantern

Lysichiton americanus

ARACEAE

Slowly, slowly, the return to life...
I am...somehow different, somehow re-created
Out of the fertile dark unknown....

                                              __*C. Maurine & L. Roche*

Botanical Description

A hairless and fleshy plant, six inches to four feet tall. The inflorescence consists of a bright yellow spathe surrounding a yellow spike or spadix. Flowers are small, yellow-green, and packed tightly on the club-like spadix. This plant has the largest leaves of any northwest native plant.

Range

Found from Alaska south to northern California, and east into northern Idaho and northwest Montana. Most common west of the Cascades. Grows below 4,000 ft.

Habitat and Ecology

Swamp Lantern flourishes in areas of mud and muck where human footsteps sink. Known also as Skunk Cabbage, it dwells in marshes, swamps, seepages, and very wet meadows. Its abode is one where worlds meet, mix and change form; water merges with earth, life with decay, leaves become humus—all around is a cauldron of transformation.

While winter still mingles with spring, Swamp Lantern's lemony spathes begin to thrust upwards through dark layers of fermenting debris. Nearby are Horsetail, Stinging Nettles, and perhaps the first forbidding shoots of Devil's Club. Sparrows and wrens—but few others—easily navigate here. Western Red Cedar and Alder often form the upper story of the community. Sunlight may only penetrate a small part of the day where Swamp Lantern grows.

Gesture

This plant's sunny yellow spathes are a welcome discovery as winter dissipates. In the dim surroundings their light is unexpectedly sublime. The spathe forms a bright cloak curving around the scepter-like spadix with a regal flourish. The genus name, *Lysichiton*, means "loose tunic", and offers a visual image for this unusual inflorescence.

The actual flowers of Swamp Lantern are its only subtle feature. Yellow-green in color, hundreds are packed on each spadix. A lens is required to view the four stamens and four tepals of each flower.

The leaves, on the other hand, contribute to the overall conspicuous quality of the plant. Following the appearance of the spathes, they emerge from fleshy underground stalks and soon overtake the inflorescence in size, reaching up to five feet long and two feet wide! Their size is unsurpassed in the Pacific Northwest region[1], perhaps in the entire temperate zone. In mid to late summer it is a humbling experience to walk among these glossy giants in the ancient forests of the western Cascade valleys. Leaves have a network of prominent veins emerging from a strong central vein, making this plant an exception among the normally parallel-veined monocots.

Among other obtrusive qualities, perhaps Swamp Lantern's aroma is most impressionable. (Unless, that is, one has the misfortune to taste the plant—a misguided and dangerous venture—in which case calcium oxalate crystals produce intense burning in the mouth.) While picking or crushing the plant produces a putrid odor, its characteristic scent is pungent and stimulating, heady with springtime fecundity. The plant actually has a repertoire of scents, emitting signals that vary from sweet to fetid as temperatures fluctuate; thereby, it targets the pollinators of the moment, whether sugar-lovers or decomposing flesh-eaters.[2] Swamp Lantern can increase its temperature, making its aromas more volatile. Through this process, it can also melt snow.[3]

Leaves and roots of this plant are food for bear, elk, and deer, and they have provided end of winter survival food for Northwest native people.

---

1   Clark

2   Mathews

3   Ross, Chambers

Boiling, with repeated changes of cooking water, breaks down and flushes calcium oxalate, rendering them safe and palatable. The plant has been revered as an ancient one with potent medicine, and was used to induce labor.[4] Present day herbalists use Swamp Lantern to relieve painful spasms and cramping, whether abdominal or bronchial.[5]

Medicine Story

Swamp Lantern is an edge-dweller; we find it at the interface of winter and spring, water and land, life and death. As with its fragrances, it possesses both alluring and repugnant qualities. It is a joyful harbinger of spring, while exuding a hint of the nether world.

Its inflorescence is both unusual and evocative, stirring the imagination. If it is a cloak and scepter, they would belong to the ruler of the underworld; if it is a lamp, it would be for guiding us within and from his domain. Its fleshy, earthy, and phallic qualities mirror our taboos and hidden powers and fears. Unlike other inflorescences whose stems hold them aloft, the bright spathe appears directly from the Earth. Its color and rapid emergence uplift our spirits along with spring sunshine and birdsong. Yet, rather than mirroring the sun in the sky as does a sunflower in midsummer, the spathe carries the inner fire of the Earth, more like the light of transmutation within a sealed crucible.

Swamp Lantern's habitat is generally not welcoming to humans. Its abode holds mysteries of life and death, dissolution and creation. Here hidden processes fuel regenerative power. In a patch of Swamp Lanterns, whole, perfectly formed spathes and spadixes are rare. The plants themselves succumb to decay and predation, for their environment teems with processes of disintegration.

Allow Swamp Lantern to illuminate what is hidden and disowned,

---

4   Mathews

5   Moore

to bring you to face the unfamiliar and perhaps repugnant. An encounter with Pluto, the ruler of the dark underworld, can feel like an assault or an abduction, whether it occurs as an event or as the undertow of our own emotional intensity. Such power dissolves our defenses and sense of self. In nature, dissolution yields fertile, life-giving soil and nutrients. We are becoming aware of the indispensable life-force and creative powers of swamps and wetlands. These same Earth mysteries reside in the fluids and processes of our physical and emotional bodies.

At times, an old self must die to allow a more whole and vital self to emerge. Innocence becomes wisdom. Swamp Lantern can guide you out of the dark winter you have believed would never end. It is the light that we carry as a gift once we have known the darkness, a light that can only be claimed in the underworld.

Red-flowering Currant

Ribes sanguineum

GROSSULARIACEAE

[God] speaks to you through your own desires,
which are the instincts of your Soul.

_Edward Bach

Botanical Description

A loosely branched, erect shrub reaching 10 feet tall. Leaves are simple, alternate, deciduous. They are palmately lobed and serrated, dark green above, paler and velvety below. Flowers are small, red to pink, and tubular, borne in a drooping raceme. Fruits are dark blue, pea-sized berries covered with white waxy bloom. The fruits are edible but unpalatable.

Range

Grows from the Pacific coast to the Cascade Mountains in low to middle elevations, across Southern British Columbia south to California's Bay Area.

Habitat and Ecology

The vivid blossoms of the Red-flowering Currant appear with the first surge of spring flowers, typically a March occurrence for the Puget Sound lowlands. While Trilliums and Salmonberry flowers gather in the forests, Red-flowering Currant bursts flame-like from sunny bluffs and shorelines. These shrubs can continue their display for over a month; meanwhile, their blossoming progresses inland and to higher altitudes. By late April or May, their flowers open in the Cascade foothills and in the western Coast Range. At Saddleback Mountain in northwestern Oregon, they join Pink Fawn Lilies to fill open woodlands with the colors of sunrise.

Look for *Ribes sanguineum* at forest edges and openings, in clearings and disturbed areas, along roads, and growing from exposed hillsides. Adapting well to rocky, dry terrain and enjoying locations where sunlight illuminates at least part of the day, they are found in a variety of settings. Characteristically, these bushes are dispersed through their habitats rather than growing together in thickets. Bearing clusters of glowing blossoms, each bush is a standout in the spring landscape. In the western part of its range, near the Pacific coast, Red-flowering Currant is most abundant, and many bushes may be scattered over a given area.

The Rufous Hummingbird migrates from Mexico and the Gulf States at winter's end, and its arrival in the Northwest coincides with the flowering of the Salmonberry and the Red-flowering Currant. The nectars of both these flowers are among the bird's most important food sources. The season's first sightings of Rufous hummingbird are often near a Red-flowering Currant bush in full blossom.

Gesture

The spectacle of this shrub in flower is as irresistible to humans as to its pollinators. The sepals make up the most colorful portion of the inflorescence. They first appear as a deep red developing from twig tips at winter's end. Bud clusters brighten and fatten as they form elongated racemes that hinge downward. These floral pendants curve when one by one flowers appear to spring open, a descending cascade of stars. The brilliance of the five sepals is accentuated by petals that are small and pearl white, closely circling five creamy anthers. A cutout star shape forms in the center, and stars seem to twirl inside of each other. Sepals fuse above the ovary to form a slender trumpet. The hummingbird's desire for sweet nectar is well accommodated while it dusts the sticky stigmas with pollen carried from other flowers.

Like crinkled hands, leaves fan open from pleated folds, emerging above the floral racemes. Leaves are five-lobed; focus on any one lobe and five more appear, revealing each part's encapsulation of the whole. In autumn, leaf tips turn brilliant red, continuing the floral gesture of vibrant color.

The Red-flowering Currant carries the message of the pentagram, pointing to nature's inexhaustible regenerative power.[1] (See chapter 3.) The

---

1   Schneider

force of regeneration and self-replication is encoded in the geometry of the flowers and leaves and emphasized throughout the plant. Branch cuttings vigorously take root in sand. Both the flowers' color and spicy to pungent scent quicken the senses, stirring the life force to renewed activity.

The sun brings out hotter hues in the flowers, while shadier locations soften the color, producing lighter pinks. The species' name, *sanguineum*, means blood red, and applies to the color of the buds as they form. The name captures the striking, activating quality of the color. By the time the inflorescence opens, its earthy red has brightened to a more vivid hue. Described as deep rose, carmine, or claret wine, it is the color of desire! In full flower, this bush pulls us into its earthy and ethereal dance.

Medicine Story

Attraction is the force that fuels nature's self-renewing pageant of boundless creativity. In desire for the other is the experience of the connection of all things. Together the Rufous Hummingbird and the Red-flowering Currant flower express life's longing for itself. They have co-evolved, exchanging information and synchronizing their rhythms. The colors and scents of one life form trigger secretions and sensations in another. The outer world enters the inner through the senses, lines of separation dissolve, and a greater system comes to light. Nature's vibrant and intricate web of desire is life-enhancing; it is held in a matrix of balance and harmony.

Modern culture, religious dogma, and industrialization have caused suppression of our natural desires; at the same time the marketplace artificially manipulates them. When we are cut off from the natural world, we are deprived of its sensory stimulation, its life-giving powers of attraction. Without this guidance we are lured by what is synthetic and degraded, and pulled further off balance. This feeds the deception that our desires are the problem, prompting more suppression and a vicious cycle. We are left estranged from our own bodies. In the natural world creatures depend on sensory cues for such things as seasonal adaptation. For example, the palette of colors of a particular season stimulates animals and humans to secrete hormones that promote processes and behaviors necessary for survival in that season.[2]

The Red-flowering Currant blossoms play an intricate and crucial role within their spring habitats. Through our bodies we have a means of approaching this mystery. As we succumb to the blossoms' magnetism, we notice their impact on our senses. How does this, in turn, kindle our appetites, body processes, sensations, emotions, and longings? Allowing and noticing in this way increases our sensitivity and awareness; we comprehend spring's renewal through participation.

The fiery song of Red-flowering Currant calls us back to the hungers of the body to recognize their sacredness, their goodness. Healing begins when we feed our senses at nature's table, allowing ourselves to be guided by our attractions. We return to community in this way, as the object of our desire has also called out to us. In responding, mutual enhancement of life occurs—we become the dance of the hummingbird and the Currant flower. In the colors, sounds, and shapes of our natural environment, our bodies' systems are prompted towards optimal health.

This physical process is inseparable from our emotional and spiritual health. The rhythms of our natural surroundings entrain our beings on all levels; we are drawn into the dance of creation. As we simply move with what delights us, our value and purpose become obvious, and our contribution gracefully unfolds.

When we suppress or cannot hear our true desires, when we try to fit ourselves into artificial containers, our life-force energy becomes depressed. The medicine of Red-flowering Currant burns through depression by catalyzing desire; it draws us into our true natures. Our deepest desires are also the world's desire, for the world longs for life in its fullest vibrancy.

---

2 Liberman

Western Trillium

Trillium ovatum

LILIACEAE

Yet mark'd I where the bolt of Cupid fell:
It fell upon a little western flower,
Before milk-white, now purple with love's wound...
> —*William Shakespeare, "A Midsummer Night's Dream"*
> *(quoted in Clark)*

> **Botanical Description**
>
> One terminal flower blooms from a single stem. Below the flower is a whorl of 3 large leaves. Flowers have 3 green sepals and 3 white petals that age to pink and lavender. Six stamens have anthers with bright yellow pollen; stigma is 3-pronged.
>
> **Range**
>
> Found on both sides of the Cascades in Oregon and Washington. Grows from the coastal redwoods in California north to the coast mountains of southern British Columbia. Also grows across the northern half of Idaho and in the Rocky Mountains from Colorado north to Alberta.

Habitat and Ecology

In the Puget Sound lowlands, the first Western Trillium flowers coincide with the Robins' first songs. Also known as Wake Robin, it may be that the flower wakes the bird, or perhaps the other way around.[1] Whatever the case, some late winter mystery cues Western Trillium and Robin to enter together the symphony of forest awakening. Winter Wrens, Bewick's Wrens, and Song Sparrows are already singing, and the forest is laced with the blossoms of Osoberry, whose new leaflets glow like green flames. Waterleaf has spread its emerald carpet, and Coltsfoot and Salmonberry have just begun to bloom.

While late February can be the time of first Trillium discoveries in the Puget Sound, the flower appears much later in higher elevations. Extending from sea level to the upper forest zone, Western Trillium is always among its habitat's first wave of blooms. This means by May and

---

1   Research shows that bird song may impact and even be necessary for the opening of blossoms. According to accounts, flower buds remained unopened in some orchards where heavy pesticide use had eradicated birds. After broadcasting recordings of bird songs in these orchards, the buds opened.

June it blossoms with a cascade of spring flowers that appear all at once as the mountain snow recedes.

Established forests, shady meadows, and stream banks provide ideal habitat for Western Trillium. The plant thrives in moist conditions, in deep, rich soil. After a wet winter, throughout the understory of a healthy forest, clusters of Trillium blossoms gleam like fallen stars. In a season of scarce rainfall, only a scattering of flowers can be found. Seed distribution is assisted by ants, who feed on the oils that congeal seeds into sticky clumps.[2] Seeds require two winters' freeze in order to sprout,[3] and the plant's tubers are dependent on a full season of photosynthesis provided by the large leaves.[4] Needless to say, picking, bulldozing, and other human disturbances are dwindling Western Trillium's numbers to precious few.

Gesture

Western Trillium is an elegant dancer twirling folds of green, its starlight flower dispelling winter's darkness. As its name implies (Trillium means triple), the number three is everywhere in Trillium's gesture. Three petals are distinct in size and color from three green sepals. The blossom opens from a long pedicel above a whorl of three large oval leaves. The stigma is divided in three. The flower is also threefold in its distinctive process of color change; a white, colored, then dark phase mark the initial, middle, and final stages of its development.

The blossoms begin the purest of white, highlighted by bright yellow anthers. After pollination, a streak of rose bleeds through each petal's central vein. This rosy streak diffuses across the petal, almost imperceptibly at first, and white becomes mother of pearl. A shimmer of pink builds to deeper blush. What better way to spend spring days than to witness

---

2   Pojar

3   Mathews

4   Moore

this color progression, on to lavender, deepening to burgundy, eventually reaching the velvety darkness of the flower's final transformation? Along the way, rain drops may add their artistry, streaking and splotching petals with unique color patterns.

The fleshy tubers of this Trillium are as much as a foot deep in the soil, with small roots extending downward. New shoots grow rapidly, piercing layers of decaying Maple and Alder leaves. Western Trillium's growth cycle is accelerated; soon it overtakes the Waterleaf and other eudicots. When temperatures are mild, it may bloom within a week from its shoot's first appearance. Should frosty times return, the leaves fold into an insulating blanket, encircling the blossom with layers of leaf and air. As the flower develops, its fragrance grows, and leaf and petal edges curve into graceful ruffles.

Herbalists make a tincture of *Trillium ovatum*, using fresh whole plants. This medicine is primarily used to quell uterine bleeding in many situations: in the case of birth hemorrhage, fibroid bleeding, heavy menses, and mid-cycle spotting. It is used to tonify the uterus and soothe irritations.[5] Historically, Native people boiled the tubers of Western Trillium for use as a love potion, and in this capacity the plant is featured in Native stories.[6]

Medicine Story

Western Trillium's threefold patterns and three phases of floral color transformation reflect nature's growth cycle. All natural processes proceed through an initial growing phase; a mature stage of solidifying and ripening; and a time of dissolution and transformation. Earth-based traditions have recognized and held sacred these three stages associated with the wisdom of the Goddess.[7] The stages are apparent in a woman's archetypal

---

5 *Ibid.*
6 McIntyre
7 Pogacnik

journey from maiden to mother to crone. Our culture short circuits this sacred threefold process. For example, we exalt youth, are ambivalent about maturity, and denigrate old age and dying. A forest is valued only for its middle phase of mature trees. The development of new trees in a natural forest is neither considered in a connected way nor valued in its own right. Mature trees are harvested, cutting off the land from the vital forces of decay. From nature's holistic view, there is no death, only transformative processes feeding new unfolding.

Western Trillium's threefold gesture expresses the cycles of growth and evolution. Humans have called upon this plant for assistance with life mysteries and sacraments: the womb and women's reproductive cycles, birth, blood, wounds, and love and infatuation. In its striking color change, Western Trillium's flower evokes the soul's journey through a lifetime. We begin shining with the light of resurrection. The world imbues us with experience, and we are deepened through wounds and passions. Finally, we surrender to transformation. Smaller cycles of birth, growth, and dissolution continually unfold within this larger cycle. These three phases correspond to the dynamic, fixed, and mutable qualities encoded in the wisdom of the zodiac. All of our creations and undertakings, like living beings, have a beginning, middle, and end. Three is a magic number pointing to the *story* of life and all things that arise within it. Appearing repeatedly in spiritual symbolism, mythic tales, and folk wisdom, three represents unity, harmony, and completion; it is the resolution of opposing forces through a higher synthesis. Three is unshakable strength and stability, the magic number that seals and ensures the blessing because it embraces and integrates two polarities.

As the stories of our lives unfold, we may notice that we have resistance to one or more of the three phases. We may lack incentive for beginnings, have trouble sustaining middles, or fear and avoid endings. Western Trillium flowers delight and refresh us at winter's end, helping us to begin anew. Their opening is a cherished event, for they bring the vitality and light of springtime, a world made new through light's return. Soon the blossom's pure white blushes and deepens with spring's unfurling; it falls from innocence into delicious color. Our awakening sweeps us once again into the currents of our life stories. Western Trillium reflects the fluidity and connection of the unfolding journey. Highlighting the sacred gifts of each phase, its flower is a lovely emblem of harmony and wholeness.

Columbia Desert Parsley

Lomatium columbianum

APIACEAE

Whatever came from Being is caught up in being,
drunkenly forgetting the way back

—Rumi

Botanical Description

Rich purple, parsley-like flowers grow from tall stalks, up to 3 feet high. Billowy blue-green foliage is fern-like; plants grow to 2 feet wide.

Range

Endemic to the east side of the Columbia River Gorge and north through Yakima County.

Habitat and Ecology

Columbia Desert Parsley is one of fifteen endemics of the Columbia River Gorge.[1] It dwells throughout the low elevations of the Gorge's Pine-Oak habitat. Here it joins the earliest flowers in bringing splashes of color to hills still dark with winter. Its blooms come on the heels of the first Grass Widows, and they are accompanied by Prairie Stars, Gold Stars, and Salt and Pepper Lomatium. The first *columbianum* flowers may appear as early as late February along south-facing road cuts or emerging from cracks of sun-warmed basalt. The flower shoots are usually the first part of the plant to develop, so look for blushes of pink amongst previous seasons' woody, brown-gray stalks. Rapidly unfurling, Columbia Desert Parsley's regal flower stalks and billowy foliage soon swell over hillsides and boulder fields. The blue-green clouds of foliage are still vibrant when Balsamroot and Lupines bloom, and, in more northern exposures, some *columbianum* flowers may still accompany these later blooming perennials.

Gesture

In full flower, the Columbia Desert Parsley is a crown jewel; its striking size, color, and textures, along with its endemic status, make it the Columbia Gorge's undisputed queen. Ranging from dusty rose to wine to

---

1   Jolley

deep rich purple, its flower's color is remarkable for its genus and family. Flowers of non-distinctive color are more typical of the Apiaceae family; many are off white to pale green. Yellows are also common, and many Lomatiums, such as Gray's Desert Parsley, brighten their surroundings.

The color of *Lomatium columbianum* speaks to the heart, speaks of passion and the rich music of life. Flowers form in compound umbells; that is, many stems radiate from a single point, then from each stem tip smaller stems grow, producing multiple miniature bouquets. Flowers seem to explode like purple and pink fireworks, lacy and light-filled. Contrasting yellow filaments reach beyond the corollas, extending the burst of light and color. Long hollow stems, themselves tinged with purple, raise *columbianum* flowers like mauve umbrellas in the spring landscape.

The air element is noticeable in the floral structure, the stalk's hollowness, and the foliage. Compound leaves unfold from purple sheaths; long, delicately thin leaflets radiate from each petiole, then divide again like fine filigree. Together they form billowy clouds. Like the flowers, the leaves have a distinctive color. Their bluish cast sets them apart from other foliage. Misty and intricate, the *columbianum* communicates with texture, color, and fragrance. Volatile oils disperse through the sun's warmth or when the plant is crushed; plant material dissolves into pungent yet gentle scent. In contrast, a large starchy taproot grounds the plant solidly and deeply into the Earth. Solidity is also expressed in the woody persistence of old stalks and large umbrella-like seed heads. Contracted and dry, accumulating over the years, the strength and rigidity of these clumps contrast with the light, expansive quality of leaves and flowers.

Medicine Story

While the Sunflower is an emblem for the Greek deity Apollo, the god Dionysus is often pictured holding a flower from the Apiaceae family.[2]

---

2   Kreisberg

When we want to "keep it together", we invoke the god of reason, order, and daylight, whose power, like the the mandala of a Sunflower, draws us into center, helping us maintain our individual integrity. Dionysus, the god of ecstasy, wild revelry, and things incomprehensible to the light of day, has other purposes.

Following the archetype of its family, the Columbia Desert Parsley's gesture radiates outward from center. From one emerges many. Leaves and flowers divide and divide again, expanding, dispersing, mingling with sunlight and air as do the volatile oils. Through orderly progressions the plant moves toward complexity and dissolution; we attempt to follow the pattern, eventually letting go, releasing the mind to spaciousness. Enclosures and interior spaces surrender to wide open accessibility, a feast spread for all pollinators. The *columbianum* plays with the edges between one thing and another; at the same time it is anchored by a strong taproot. The tough stalks that persist through the continual change of the growth cycle point to something changeless. Through the seasons, they mark where each year form uncoils from formlessness.

Before winter's end the Columbia Desert Parsley's vibrancy and eagerness for life catapult it into rapid growth, the rich purple of its flower overflowing onto stalks and sheaths. As it expands, forms and colors differentiate, breaking the monotone of winter. The forces of resurrection color this plant and set it in motion. With it awakens life's longing for itself. The *columbianum*'s medicine is passion for life. The plant engages all the senses, our doorways of communion. We breathe deeply, opening the heart, dissolving edges. Following the wine-colored expanding blossom, we melt like a nature mystic into the ever-shifting play of the divine. The plant teaches us to be strongly and deeply rooted so we can surrender to the spell of nature's infinite creativity.

Poet's Shooting Star

Dodecatheon poeticum

PRIMULACEAE

The soul that rises with us, our life's Star,
Hath had elsewhere its setting,
And cometh from afar:
Not in entire forgetfulness,
And not in utter nakedness,
But trailing clouds of glory do we come

_William Wordsworth

> **Botanical Description**
>
> Flowers have five reflexed, bright, pinkish-purple petals with a yellow and red-ringed tubular base. Stamen filaments are united at the base into a purplish tube. 2-10 flowers grow from 6-12 inch stems. Leaves have serrated edges and grow in a basal rosette. Leaves and stem are covered with fine hairs.
>
> **Range**
>
> Found only in the mid to eastern Columbia River Gorge.

Habitat and Ecology

The Poet's Shooting Star is one of fifteen wildflower species that grow only in the unique habitats of Washington and Oregon's Columbia River Gorge.[1] This endemic Shooting Star is found in lower elevations in the Pine-Oak belt of the Gorge. Summer means intense sun and moisture-leaching winds in this dry, thin-soiled habitat. Early to mid spring, however, brings abundant wetness. Grass Widows, Yellow Bells, Gold Stars, and some Lomatiums are already in full glory when these magenta Shooting Stars begin to dot grassy meadows and open woodlands. They herald the second wave of early blossoms in this area. Joining them are Camas, Prairie Stars, Upland Larkspur, Blue-Eyed Mary, and Balsamroot. Poet's Shooting Stars gather by springs and seeps and along basalt outcroppings. After a wet winter they form impressive clusters of wind-tossed pink.

Gesture

The flowers of the genus *Dodecatheon* seem to spring from human imagination. Could they be miniature rocket ships descending earthwards with dynamic precision? *Dodecatheon* comes from the Greek, meaning

---

1   Jolley

"twelve gods". Perhaps the flower invoked images of heavenly emissaries for its botanical classifiers.

The *poeticum* has five smooth and shiny petals that stream backwards, exposing a deep purple "nose cone" ringed with yellow and red wave patterns. From here stamens converge in a purple filament tube, and the thin, maroon style extends beyond the stamens forming a precise point. The magenta petal color of the Poet's Shooting Star intensifies just before meeting a bright yellow base.

Viewed up close, this flower resembles a meteor or sky rocket. Streaking earthward, it arrives among oak-scattered hills to release the spell of winter. Along with the dynamism of its colors and form, the movement of the blossom through its cycle is striking and distinctive. The closed bud points upward to the sky. Soon it begins to bend, moving through a horizontal orientation until it is focused straight down. In this position the flower opens, aiming precisely earthward. Here it hovers from its delicately arched stem while bees approach from underneath, bouncing as they gather nectar. After pollination, as petals fade, the flower begins its ascent as its long stem straightens. The flower again points skyward in its final transformation to seed.

The Poet's can be distinguished from other Shooting Stars by its slightly toothed leaf edges. The only other Shooting Star with serrated leaves is the White Shooting Star (*D. dentatum*), whose striking black-accented white flower sets it apart. The Poet's leaves are basal, elongated, oval, pointing upwards. They are soft with tiny hairs. Stems, also downy with hairs, are wiry and long, branching at the tips. Each branch bears a flower.

Medicine Story

Shooting Star flowers conjure images of celestial visitors, of pink starlight that showers the land. Perhaps they are seeding new ideas, sowing wisdom from the cosmos, bridging us to dimensions beyond the watery membrane of the Earth.

Dodecatheon imitates the human journey. We are formed from the humus of the Earth, yet we seem to belong to a larger universe, to something limitless. Our souls require that we live this paradox. We must immerse ourselves in Earthly life while remembering that some transcendent purpose has spun our lives in motion. How easy to forget that we are connected to spiraling galaxies. Powerful forces swirl outside our grasp, and we lose touch with any sense of mission for this life on Earth. We are susceptible to depression, boredom, perhaps self-deprecation, as the vast mystery of our existence eludes us.

The gentle and dynamic Shooting Star can assist us when we slog through daily routine, consumed with longing for elusive visions. Early spring brings the opportunity for pilgrimage to the Poet's Shooting Star, whose first appearances are in the Catherine Creek area, on the Washington side of the Gorge. Here we find hillsides strewn with pink, and the flowers speak to us in the language of the cycles of the soul, the connection between Earth and cosmos. We catch our reflection in magenta petals and something yet unrealized swirls within. Each small flower, humble and exotic, grounds a unique droplet of beauty carried from afar.

The Poet's Shooting Star makes its home solely where countless factors have converged into one precise life-giving ecology. This one-of-a-kind habitat is a microcosm of the entire Columbia Gorge area, where myriad life opportunities are created as a great river winds directly through the Cascade Mountain Range. The Gorge, in turn, is a microcosm of the near impossibility of our one precious planet, in which is reflected and held each irreplaceable life.

Common Camas

Camassia quamash

LILIACEAE

There was a time when our people covered the whole land, as the waves of a wind-ruffled sea cover its shell-paved floor. ...At night when the streets of your cities and villages shall be silent, and you think them deserted, they will throng with the returning hosts that once filled and still love this beautiful land.

__Chief Sealth

> **Botanical Description**
>
> One to 2 foot tall erect stems; leaves basal and grass-like. Flower spikes bear clusters of many flowers, each with 6 blue to purple tepals, 6 stamens, and a single pistil with green ovary.
>
> **Range**
>
> Grows from southern British Columbia to northern California, extending eastward to the intermountain west. Found in the southeast part of Vancouver Island and a little bit in Alaska.

Habitat and Ecology

At the height of the spring bloom, undisturbed lands with ample moisture are awash with the blue-purple blossoms of the Common Camas. Waves of fresh color sweep through areas that are wet in spring but baked dry by summer, including the prairies south of Puget Sound, hillsides and outcroppings of the Columbia Gorge, wetter pockets of shrub-steppe lands, and expansive meadows near the Rocky Mountains. Across its extensive range, Camas once covered many thousands of acres.

Camas is deeply intertwined with humans and it has been impacted by human history. The bulbs of the Common Camas were a vital source of food for many Native tribes, including the Nez Perce, who named the plant *Quamash*, meaning "sweet".[1] Traditionally, long roasting in pits covered with leaves and grasses broke down the inulin in the bulbs, rendering them good-tasting, digestible, and sweet. Native people knew how to cook the bulbs for flavor and nourishment, and knew how to enhance and help perpetuate their growth. The relationship of Northwest Natives to the Camas connected people within community, with ancestors, the land, and all that grew from the land. Each spring or early summer tribes journeyed to meadows that had been filled with Camas for thousands of years. Some

---

1   Carville

tribes would tend the Camas fields; sometimes a family would work in the same patch of Camas for generations. This involved year round work of clearing rocks and weeds and burning to abate encroaching brush and conifers. Only larger bulbs were harvested, ensuring the continual development of young plants.[2] All of the activities of tending, harvesting, preserving, cooking, and feasting were grounded in awareness of the sacred wheel of life. The people held the knowledge that all things are connected, together participating in life's eternal return. Camas brought the whole community together in shared labor celebrated in festivals, held within a matrix of ceremony, reverence, and reciprocity. Native people were part of the ecology of the Camas meadows. Their interventions were life-giving and sustainable through many generations.

With the coming of the white settlers in the late 1800s, the vast Camas fields began to diminish, for these settlers plowed the land and brought many grazing animals as well as pigs, who uprooted and feasted on the Camas bulbs.[3] Settlers suppressed wildfires, so crucial to the creation and preservation of Camas habitats. The long-term balance of land, people, and plants was suddenly disrupted.

Camas has been an integral part of communities—including Native peoples—that today are rare and endangered or no longer exist. For example, the glacier-carved prairies south of Puget Sound, where several Native tribes once dug Camas bulbs, and where many species of birds, butterflies, animals, and plants are now endangered, is only three percent of its original size.[4] Development continues here, its machinery rupturing the land's living matrix, carving the sacred wheel, immune to the breath-taking beauty of Camas.

---

2   Pojar et al.

3   Carville; Mathews

4   Nature Conservancy

Gesture

Each Camas flower is lovely onto itself. Both sepals and petals are the same color and shape. Thus referred to as "tepals", they amplify the display of color. Slender and washed with shades of purple to blue, tepals extend into a six-pointed star. In the star's center is a lime green ovary with a lavender style and three-lobed stigma. Six yellow anthers twirl bright accents from long filaments. This starry splendor is magnified in the terminal clusters of flowers that open all at once from the plant's upright stem. Each plant, in turn, joins with many others, sometimes thousands, and together they become a dazzling expanse of color. Rippled by spring breezes, Camas communities have been compared to—or even mistaken for—sparkling lakes of clear water when viewed from a distance. This effect is especially pronounced east of the Cascade Mountains and towards the Rockies where the more blue to turquoise *quamash* varieties grow; varieties west of the Cascades are more purple.

Though it can be found in smaller clumps, Camas is most known for painting the landscape with great swaths of color. One can become lost in a single flower, as well as be impacted by dramatic displays of many plants. In Camas' gesture smaller parts combine to form something of a greater magnitude. Coming upon a large meadow in full blossom, one can only stop everything, breathing deeply.

Medicine Story

A field of Camas in bloom brings a feeling of expansiveness, a sense of possibility and inspiration. The mantle of blue-violet, a color of higher frequency, quickens and uplifts, stimulating vision and intuition. Our vision opens to a broad, inclusive perspective. Camas' gesture leads our focus through the details to behold the bigger picture. Each flower and each plant repeat the theme of individual parts connected to something greater. Camas' six-fold geometry, vertical flower spikes, and relationship to water all hint at its role as a conduit between the lower and upper worlds. (See Chapter 3.)

Camas' gesture reveals the relationship of the microcosm to the macrocosm. Each aspect of the plant repeats, "As above, so below". This wisdom is expressed in Camas' relationship with Northwest Native Peoples, who, like the Camas, once flourished in these lands, their lives interwoven in many ecosystems. The people who depended on Camas for their survival lived with conscious knowledge that they were embedded in communities within greater systems. Individuals were held in a matrix that included the tribe, ancestors, those yet unborn, the land with its living systems, and the land's history. Native people and Camas formed a great connected community, which was honored through sacred ceremonies. Yearly tasks were carried out with prayer, gratitude, and reverence, acknowledging all the relationships and the big embrace of Mother Earth. Because Native people were conscious of the big picture, they were aware of the laws of nature, including eternal return and reciprocity. Keeping and holding sacred the laws of nature ensured that their activities were sustainable and life-enhancing.[5]

Camas holds the wisdom story of the Indigenous perspective. This plant offers medicine to modern people who have been long separated from land, community, and ancestors—from our human essence. Modern culture perceives the world in unrelated pieces; therefore, challenges, aspirations, and activities are not considered in greater contexts. Our behaviors have far-reaching consequences. We must consider our actions in terms of an ecosystem, a bioregion, a planet, and in terms of unborn generations. Correspondingly, we shed light on a problem through the larger perspective of its meaning within a family, community, or nation, and in the context of past history. Individual symptoms reflect global disharmony and point to a path of healing the greater community.

To honor the call of Camas, begin with a personal dilemma and ask how this might be related to the history of the land on which you are

---

5   LaDuke

living. Following one small strand can lead to the recovery of the whole. Camas medicine prompts us to see from as big a perspective as possible. We often feel like failures when we cannot individually, heroically resolve our problems. Our struggles are both personal and collective in nature. Our pain calls attention to unresolved suffering and imbalance within our families, ancestral lines, communities, and within our web of planetary relations. The very land we live and work on has its own history that includes oppression, violation, and the suffering of many beings. This has never been reconciled, and this disharmony underlies our daily activities. The spirit of Camas guides us to the stories held in the land. Their revelation is needed for healing to begin and to restore relationships that are life-giving and sweet.

Bitterroot

Lewisia rediviva

PORTULACACEAE

...Where there is hatred let me sow love....
Where there is despair, hope,
Where there is darkness, light, and where there is sadness, joy....
For it is in giving that we receive—
...in dying that we are born again.

__St. Francis of Assisi

Botanical Description

Flowers pink to rose, occasionally white, up to 2 inches in diameter, with 10-18 spreading petals, 40-50 stamens, and 6-8 stigmas. Basal rosette of fleshy, linear leaves withers before blossoming.

Range

In Washington: grows from the Cascade Crest eastward, from the lowlands to the alpine. **Other areas:** found from southern British Columbia to southern California, from the Coastal Ranges in California eastward to the Rocky Mountains and the intermountain west; grows throughout the Sierras from 200 to 9000 ft.

Habitat and Ecology

Bitterroot makes its home in harsh, barren-looking environments. It is found in lithosol (very dry "rock-soil"), shrub-steppe, and subalpine areas. Blooming close to the ground, the softly luminous cups of the many petaled flowers appear to float from rocky outcroppings, ledges, and scree slopes. Growing from many types of rock, including basalt, granite, and limestone, the flowers resemble mysterious water lilies.

Bitterroot survives the intense summer heat of its habitats by going dormant after blooming. By late spring or early summer all evidence of these ethereal rock lotuses has vanished. When the rains of late summer or early fall refresh the land, succulent, pencil-thin leaves sprout from fleshy rhizomes. These basal rosettes of rich green endure through the winter. They may not be noticed until spring's earliest flowers draw attention to the land. At this time the leaf rosettes are a prelude to the Bitterroot blossom, which will appear about a month after the first spring ephemerals, coinciding with the flowers of Balsamroot and Lupine.

Gesture

The species' name *rediviva* means "reborn" or "restored to life". There are several explanations of the origins of this name. Some say it refers to the plant's traditionally recognized medicinal qualities.[1] Others tell the story of the Bitterroot specimen that Meriwether Lewis gathered and pressed completely dry. The dried specimen was brought back to Pennsylvania, and about two years later it showed signs of life and was planted. It then grew and blossomed.[2] In truth, the name works on many levels, for restoration to life is part of the core essence of *Lewisia rediviva*. Its light-filled pink, apricot, or deep rose flowers restore its barren wasteland habitats to vibrancy. After flowering the plant remains completely dormant until resurrected through the first rains of late summer or fall. In spring the buds look limp and lifeless; camouflaged among rocks, they are often stepped on, perhaps even broken and bent, their life force seemingly crushed. Amazingly, these buds soon burst open satiny petals and a flourish of stamens whose bright pink anthers surround twisting cream-colored stigmas. Dynamic and light filled, the flower's opening is a dramatic rebirth, unanticipated among hardpan and rock rubble. Flowers almost immediately close with evening's approach or even a passing cloud, and just as abruptly they unfurl in sunlight. These contrasts highlight the magic of the Bitterroot as it continually, unexpectedly reignites.

Bitterroot demonstrates its resurrection power in its adaptive strategy, which gives it tenacity, and the ability to survive extreme heat and dryness. The leaves produce sugar as they absorb late winter and early spring sunlight. Roots store this sugar and gather the moisture of rain and snowmelt. Later in the spring with waxing heat and scarcer rainfall, leaves wither and drop just as the flower begins to bloom. The plant is sustained through blossoming by the sugar and water stored in the root.

---

1   Blackwell

2   Clark; Turner, Gustafson

These carrot-like roots were an extremely valuable food source for Native people. Dug early in the spring before they became too fibrous, they provided restorative nutrition at winter's end, extending the theme of rebirth on another level. The roots' intensely bitter orange bark was peeled, and they were boiled and eaten or dried for storage. Just as this plant sustains its life force over long periods of time, the roots preserve their vitality and nutrition; their importance as food for Native people lay both in their availability and in this ability to keep well.[3] Europeans related differently to this plant, for even when peeled, the roots were too bitter for their taste; thus the common English name, Bitterroot, was inspired.

Medicine Story

The humble little Bitterroot blooms right next to the ground in its harsh and forbidding environments. Its lifeless looking buds lie on the ground among the stones, unsuspected of splendor. Through Bitterroot's adaptability it blends and disappears into its desiccated, unyielding habitats—places notable for hardness, the darkness of basalt, the density of granite. Yet rosettes of verdant leaves hint at rosettes of many petals, at flowers expanding into generous clusters. In contrast to density and darkness, petals are translucent, softening and uplifting with rose-colored light. Flowers bring the land alive with their luminous cups. Bitterroot flowers are the life and light hidden in the rocks, the magic in the wasteland. They are the promise of eternal return, the assurance of the regenerative power of the land. Their opening reveals the lotus in the rock, the light and color of the minerals, holy sparks hidden in matter, life in apparent death.

The medicine of Bitterroot is the ability to bring light into the darkness, to transmute density into translucence. This soft pink lotus-shaped flower can teach us how to keep our hearts open in desolate places where suffering reinforces despair and illusion. Many compassionate people

---

3   Clark

choose to serve in prisons, war zones, and ghettos, and through their ability to recognize the light hidden in the darkness, they are able to kindle luminescence in these places. These people possess the medicine of the Bitterroot, medicine that helps us recognize and amplify light and vitality in places or states of being that seem inhospitable and unyielding. Bitterroot can teach us to sustain ourselves through darkness, holding faith in life's renewal and the power of love and light to spring from unanticipated sources. The colors of this flower are the colors of the heart, of compassion. The flower's translucence and lotus-like nature tell the story of our continual reemergence from density to clarity as the heart frees us from illusions of separation.

Bleeding Heart

Dicentra formosa

FUMARIACEAE

Love gives naught of itself and takes naught but from itself. Love possesses not nor would it be possessed; for love is sufficient unto love. When you love you should not say, "God is in my heart," but rather, "I am in the heart of God."

              __*Kahlil Gibran*

Botanical Description

Bleeding Heart is 1-2 ft tall with fern-like foliage and heart-shaped flowers. Foliage continues to grow after blooming ends. Flower color varies from pale pink to deep magenta. Plants spread by seed and rhizome, and can form large patches.

Range

Grows in moist woods from lowlands to lower montane from the Cascades to the coast. Found from British Columbia south to California.

Habitat and Ecology

When spring is well underway, Bleeding Heart spreads its enchantment through broadleaf thickets, conifer forests, ravines, and streambanks. In the lowlands, this usually occurs in late March or early April. When Bleeding Hearts begin to bloom some of the earliest flowers have begun to fade, Western Trilliums are turning lavender, and the world teems with life and greenness. Bleeding Heart belongs to this time of heightened faery activity and to the realms of soft shade, moisture, and dark humus. However, in mid or even late season, after spring growth has waned, Bleeding Heart blossoms can surprise us in the lower forests of the western Cascades or as we ascend trails to the subalpine. These encounters can occur in the shelter of trees or thicket, or even along old glacial moraines. Though summer plant communities are smaller, they reveal Bleeding Heart's adaptability. Plants can survive in very wet to rather dry situations, and they spread efficiently from rhizomes. Seed distribution is assisted by ants, who devour the seeds' white, oily appendages.[1]

---

1   Pojar et al.

Gesture

Bleeding Hearts in flower call for abundant adjectives. Graceful curves of smoky rose hearts are poised above feathery blue-green foliage. The interplay of shapes and colors—three dimensional pendant hearts, finely etched leaves, colors that appear veiled with mist—conjure faery stories, romance, and magic. These plants seem to spring from the interface of human imagination and nature.

Close examination reveals that the inflated heart shape is created by two fused petals that are enclosed by a larger pair of fused petals. The larger petals each have a nectar-filled spur at the bottom, creating the effect of a droplet of blood or teardrop. The flower's soft rose curves enfold a hidden interior space, its graceful arches meeting in perfect bilateral symmetry.

Delicate and sensitive is Bleeding Heart. Spring breezes stir its feathery leaves and dangling heart clusters. The fine points of emerging leaves are exquisitely soft. As well as with air, the plant has a kinship with water, as expressed in its affinity for moist habitats, its flower's droplet-like sac of nectar, and the watery sap of its stems. Both succulent and airy, Bleeding Heart weaves its spell through edges and openings, beckoning closer communion.

Bleeding Heart's signature points to its medicinal gifts to humans. Both roots and leaves are formulated as tinctures and used internally or topically for pain relief. This medicine soothes the pain of the deep lying nerves. It can be of assistance when there is trauma in any part of the body. After an accident, violence, or other emotional shock and trauma, the tincture can calm and soothe shakiness, anxiety, fear, and anger, assisting one to return to center. It helps people who are depleted or who have struggled with illness for a long time to regain strength through reestablishing the body's anabolic (building) processes. It tonifies and stimulates, increasing the appetite, helping the life force to return.[2]

As a flower essence, Bleeding Heart assists with the emotional trauma of loss of a loved one. It eases the heart's pain and helps a bereft person to regain vitality through emotional self-nourishment. The essence is indicated for cases of over dependency and neediness, helping one to reclaim his or her spirit in order to be able to give and receive love with freedom.[3]

Medicine Story

Perhaps our enchantment with Bleeding Heart comes from a recognition of its analgesic properties, from an intuitive knowing that it is a balm for physical and emotional wounds. It is as though within the soft curves of its flower or through its sensitive leaves it can understand human pain and folly. In the presence of its fanciful charms, we forget to take ourselves seriously.

The heart shape is widely associated with matters of love and relationship, as is the human heart. The heart symbol represents the subtle level of the physical heart. One way to view a heart formation is as two golden spirals that remain separate but touch and mirror each other. Golden spirals are found everywhere in nature. (See Chapter 3.) Maintaining the proportion of the golden mean, they underlie all natural processes of regeneration and transformation. Two adjacent golden spirals can be found in the two vortexes that unfurl to become the two lobes of a heart shaped leaf, such as that of a violet. Observe the opening of a violet leaf in the springtime and you will understand why certain Earth based cultures referred to these double unfurling vortexes as "the nostrils of Pan".[4] They saw in this process the force, or vital breath, that animates nature.

Golden spirals—and vortexes, their three dimensional version—grow

---

2   Moore
3   Kaminski, Katz
4   Schneider

through self-accumulation from a still point within an infinite center. They whirl into being when opposing forces collide; they are nature's resolution of differences and resistance. Vortexes are constantly forming in moving water where it encounters obstacles and when flows of different speeds and temperatures meet. Our physical heart is enveloped by a golden spiral-shaped muscle, and the action of the heart creates a vortex of electromagnetic current.[5]

In the heart center is the resolution of differences and conflicts. The heart carries the knowledge of oneness, and with its intelligence we experience relationships as mirrors of self. The wisdom of nature reveals all beings as one multifaceted organism. The core of nature's architecture, the golden spiral, demonstrates that differences resolve into growth and transformation; oppositions feed the power of regeneration.

Two spirals together in the imagery of the heart shape convey the potential for creative transformation, unity consciousness, and compassion within the dance of relationship. In the heart shape, the spirals touch but unfold separately. Fetuses develop through spiraling vortex movement, and identical twins form mirror spirals, or a heart shape, as they grow. This heart formation occurs with other simultaneous growth processes in nature. It allows adjacent entities to unfold together while never interfering with the other's growth.[6]

These associations bring insight into the subtle and physical medicine of the Bleeding Heart. This delicate flower reflects the freedom and spaciousness inside our core. Within the heart is the realization of connection behind the illusion of separation. Here we can release attachments, find renewal, and reengage in life as more whimsical and compassionate players.

---

5   *Ibid.*
6   *Ibid.*

Calypso Orchid
Calypso bulbosa
ORCHIDACEAE

As I learn more about the seed of true self that was planted when I was born, I also learn more about the ecosystem in which I was planted—the network of communal relations in which I am called to live responsively, accountably, and joyfully with beings of every sort.

__Parker Palmer

Botanical Description

Plants are 6-8 inches tall with one shiny leaf at base of stem. Flower is pink overall with a purple slipper-like lip.

Range

Found through Canada, from B.C. to Labrador, and south to California, Colorado, and northern Arizona.

Habitat and Ecology

In mid to late spring, concealed in the mossy depths of the Pacific Northwest's mature forests, is the unexpected magic of a small, often solitary flower, the Calypso Orchid. In Homer's Odyssey, Calypso, whose name means "concealment", was a beautiful sea goddess-nymph, who lived hidden in the forest. She is encountered by the shipwrecked Ulysses when he drifts to her island. The wildly imaginative Calypso Orchid may well hold observers captive just as Ulysses was detained by the sea goddess. This flower is a portal through which the spirit of the forest moves. However, on the way from trailhead to destination these little plants are more often unnoticed than wondered at. Invisible to most are the complex circumstances and life forms that have conspired in this moment of blossoming. The ornate Calypso Orchid is actually an entire web of relationships.

This Orchid belongs to tall, mature coniferous communities, humus-rich forest floors, and deep shade with fleeting pools of illumination. Specific habitat, pollination, and germination requirements mean that this plant is a rare, co-created event. The Calypso Orchid produces thousands of minute seeds, well-suited for wind distribution, but lacking in nutrient stores. These seeds cannot germinate unless they are penetrated by the microscopic tubes of particular fungi species; these fungi convert insoluble starches in the seed to the immediate sustenance of simple sugars.[1] From the start, the Calypso Orchid is linked to the great network of tubes, or hyphae, that make up the fungus. With this liaison the plant is plugged into yet greater connections and benefits, for the fungi have extended a vast mycorrhizal net that taps into the root systems of conifers, perhaps other plants, as well. This arrangement works well for the Calypso Orchid, whose rudimentary root system and one leaf, withered by summer, cannot adequately nourish the plant. The fine threads of mycorrhiza funnel soil nutrients, minerals, nitrogen, and water to the Orchid more efficiently than plant roots. And, not to worry that the Orchid is capable of minimal photosynthesis; the mycorrhiza deliver carbohydrates produced in the sun-filled upper forest canopy. The mature conifers are indispensable to the system for photosynthesis and for providing water through their deep roots during the dry season. The fungi collaborators provide protection from other fungi and bacteria that are detrimental to plants.[2]

An enormous amount of pollen grains are required for fertilization of the Calypso Orchid's copious seeds. The plant rises to this challenge through its flower's sophisticated and ingenious structures, visual designs, secretions, and timing mechanisms. The flower is custom-designed to accommodate its few specific pollinators. It irresistibly lures these insects, maximizing cross pollination opportunities and minimizing wastage of pollen grains.

Gesture

The Calypso Orchid helps us understand the inextricable connection of a plant's gesture to its ecology. It's highly embellished, evocative flower has developed in relationship with insect pollinators. Its small, round, bulb-like corm, shallow and tenuously held in humus by thread-

---

1   Clark
2   Mathews

like roots, necessitates the plant's auxiliary web of soil-based relationships. In heavily shaded abodes, partnerships can have more advantage than leaves. The Calypso Orchid's one leaf grows directly from its corm in early fall when light and other foliage are fading. This leaf prevails through winter storms, only to shrivel away by the time the flower has gone to seed in late spring or early summer. The minimal root and leaf structures are a striking contrast to the Calypso Orchid's highly developed, intricate flower.

This unusual and evocative flower has inspired an abundance of common names. These names associate it with faeries, Aphrodite, deer heads, slippers, and nymphs. With its five narrow, rose-magenta tepals flaring and twisting above the flashy "slipper" petal, some might see a court jester. The slipper, or pollinator landing platform, glistens with translucent white, and inside and out it is streaked and splotched with pinks and earthy red. A flourish of yellow hairs embellishes the "tongue" portion of the slipper, ensuring that the pollinator will be entangled long enough to receive and deposit ample pollen. An effective structure of fused stamens and pistil, called a column, is conveniently located just above this perch. Two spurs protrude from the petal's tip, creating yet another embellishment, or perhaps the forked tongue of an exotic reptile. If the unrestrained artistry and jewel-like colors of this little flower are not enough to blissfully delay your hike, its fragrance, once sampled, will be forever irresistible.

The Calypso Orchid is the only pink Orchid of the Pacific Northwest region; it is a privilege to encounter one or two flowers or, if very fortunate in an abundantly wet year, a small grouping. Now endangered from footsteps, picking, and habitat disturbance and destruction, these plants call for extreme gentleness and respect. Picking ensures their death, as the slightest tug will tear the delicate, shallow root fibers.

Medicine Story

The Calypso Orchid's liabilities are overcome through its intricate, far-reaching relationships with its surroundings. Its complex cooperative ventures blur the lines between one organism and another. Orchids, fungi, and trees interface in one system. In its sophisticated relationship with pollinators, the flower has taken on some of their animal characteristics. Its bilateral symmetry, interior spaces, and unusual designs convey movement, sensation, emotion. We see animal forms that shift between mammal, insect, and reptile; we feel in the presence of strange enchantment.

The Calypso Orchid pulls us into dreamlike perception; the world reveals itself as fluid, interconnected. The plant embodies the light-filled forest canopy, the shady forest floor, and the dark underground realm. Within its flower, these three worlds shapeshift, evoking a swirl of moods, darkness and light.

Consider an aspect of your life from a relational, flowing, ever-shifting perspective. For example, the systems of our bodies are the activities of countless microorganisms. All that we have manifested has arisen through a web of imagination, ideas, and communication—phenomena which are more collective than belonging to any one individual. Our connectedness makes it possible to tap into other life forms, textures, ways of knowing and being in the world. The upper, middle, and underground worlds are all available to us. They reside within us, and we must honor the sacredness of all three. The medicine of Calypso Orchid brings awareness to the extensive web that sustains us, allowing us to consciously interact with it. Through this expanded perspective we are able to enlist the web in serving and enhancing the whole.

Shining Oregon Grape
Mahonia aquifolium (Berberis aquifolium)
BERBERIDACEAE

There is in all visible things...a hidden wholeness.
 __*Thomas Merton*

Botanical Description

Shrubby plant, 3 feet or taller. Compound leaves have 5-9 leaflets, glossy on top and pinnately veined. Bears clusters of bright yellow, honey-scented flowers.

Range

In Washington: Grows both east and west of the Cascades and throughout the Columbia River Gorge. **Other Areas:** Found from southern British Columbia through Oregon, into Northern California's Coast Range and on the west side of the Sierras.

Habitat and Ecology

Shining Oregon Grape, also known as Tall Oregon Grape, likes to grow in dry locations with abundant sunlight. It can be found in low to middle elevations at forest edges or in the open, often on rocky slopes or bluffs. It thrives in the rainshadow coastal areas and islands of Washington and southern British Columbia. From south and west facing sea bluffs, plants can form tall thickets, and in April and May they are a spectacle of bronzy leaves and bright yellow flower clusters.

Two other species of *Mahonia* grow in Cascadia. They are quite similar to the *aquifolium*, mainly distinguished by their size and habitats. The more common Cascade Oregon Grape (*M. nervosa*) grows only on the western side of the Cascades where it prefers shady forests. It spreads under conifers into a low, leafy cover. The small, ground hugging Creeping Oregon Grape (*M. repens*) is found only in dry areas east of the Cascades and Sierras, from low shrub steppe to rocky, exposed mountain ridges. Shining Oregon Grape is intermediary between these two other species, for it thrives in dry, open habitats, both east and west of the Cascade Range.

Gesture

The three Oregon Grape species of Cascadia are akin to each other in their overall gesture and medicinal qualities. The distinctive traits of the *aquifolium* come to light through comparison with the *nervosa* and the *repens*. In size the *aquifolium* exceeds the other two, for it can grow over six feet tall. Its leaflets are extremely shiny, so much so that they appear wet. Its botanical name, *aquifolium*, meaning "water leaf", refers to this quality. All three species have leathery, prickly-edged, leaflets. The *aquifolium* leaflets are the most prickly, just a bit less hostile than holly leaves. The leaves of the *nervosa* are longer with more leaflets, each with palmate veins; that is, three main veins originate from the same point at the base. In contrast, the leaflets of the *aquifolium* are five to nine per leaf and pinnately veined: there is one deeply inscribed central vein from which other veins branch. This conifer-like inscription echoes the taller, more upright gesture of the *aquifolium*. In all three species of Oregon Grape, leaflets remain throughout the year. When leaflets are two to four years old, vivid crimson or orange splotches begin to seep into their luxuriant green. Their brilliant designs or solid crimson may appear at any time of year, indicating they will soon fall to the ground and be replaced with new growth. The leaves of Shining Oregon Grape turn bronze-green with exposure to abundant sunlight.

The clusters of bright yellow, honey-scented flowers are most abundant in the *aquifolium*. They are as sweet and alluring as the fortress of leaves is forbidding. In the end, the sweetness of the flowers wins out, for it is worthwhile to endure scratches to take in their loveliness. Two whorls of three sepals each encircle two inner, three-petaled whorls. This creates a miniature coronet, beginning with delicate green, often with a blush of red, embracing inner layers of daffodil yellow. As spring turns to summer, bright blossoms become deep purple berries. Frosted with pale powder, known as bloom, they grow in grape-like clusters. Historically a food source, and still gathered for jelly, they are usually mixed with sweeter fruits and plenty of sugar. Their sourness is a stark contrast to the irresistible flowers that precede them.

The stems and roots of Oregon Grape are bright golden yellow due to concentrations of berberine, a bitter alkaloid. The ground bark of the lower stems and roots creates a vivid yellow dye traditionally used by Native people.[1] Because of its berberine and other alkaloids, Oregon Grape is a valued medicinal plant that has an affinity for the liver. Stems, roots, and leaves are formulated into bitter tonics that activate digestive enzymes and stimulate a sluggish liver and gall bladder. Oregon Grape root strengthens the immune system to fight infection. This plant's medicine detoxifies, clears stagnation and inflammation, is a blood purifier, enhances all liver functions, and awakens and fortifies the whole system.[2]

The flowers of the *aquifolium* species of Oregon Grape are made into a flower essence by the Flower Essence Society of California. This essence is useful for those who expect hostile and unfair treatment from others. It helps instill a sense of trust, warmth, and good will, replacing paranoia and defensiveness with positive expectations.[3]

Medicine Story

Shining Oregon Grape contains the vibrancy and warmth of spring's golden sunshine. In many ways it gives back the sun's brightness—through its shiny leaves, sparkling flowers, and the rich, bright yellow in its stems and roots. However, the sweet, alluring flowers are perched among forbidding leaves, the lovely blue berries turn out to be intensely sour with bitter aftertaste, and bitter is the plant's blessed medicine.

---

1 Pojar
2 Moore
3 Kaminski, Katz

Oregon Grape is both a friend with a sunny disposition and a prickly, harsh encounter. Bitterness resides in its fruits and alkaloids, and the spiny leaves say, "stay away, don't touch me!" Through the polarities of its gesture, the plant reminds us of human interactions, of our meetings and edges, our sense of empowerment. A prickly, harsh, and bitter disposition is one response to past hurt and trauma. An imbalance in the liver correlates to anger that, instead of being expressed in a powerful way, is stagnant, repressed, dominating, or explosive.

Anger, like bile, is needed to activate and stimulate, to awaken the inner warrior to appropriate action. Healthy, fluid, protective boundaries correlate to appropriate immune function; they derive from strong, balanced personal power. Weak, self-effacing boundaries compromise one's integrity in the same way as a weak immune response. The other extreme, constant mistrust and hair-trigger anger, is mirrored in multiple allergies and autoimmune disorders. Both extremes degrade our health and often occur together. Both are conditioned responses originating in unresolved trauma and unacknowledged violation. Comparing Oregon Grape's herbal actions with its use as a flower essence reveals the connection between physical, emotional, and psychological states, as well as past history. A "hidden wholeness" unfolds as we attend to the different levels of the plant's medicine and the polarities expressed in its gesture. Similarly, the story of our own wholeness is revealed through our symptoms, though they be disruptive, painful, and perplexing. Attending to symptoms, we find within them greater self-understanding and an opportunity to reclaim banished aspects.

Sweet and bitter are inseparable qualities of Oregon Grape; both offerings help to make us whole, and in wholeness, our authentic power shines like a radiant jewel. The plant's bitter extracts and its sunny flowers bring strength and light into our core. Its medicine helps us embrace the overall goodness of life.

Nootka Rose

Rosa nutkana

ROSACEAE

The most precious gift is that which opens us to life itself.

—Joanna Macy

> **Botanical Description**
>
> Shrubs grow in thickets and bear pink 5-petaled flowers with numerous stamens. Flowers are large—up to 4 inches across—and grow singularly from branch tips. Petal tips are broadly notched. Five green sepals have toothed tips and are the same length as petals.
>
> **Range**
>
> Grows both east and west of the Cascades, from Alaska to northern California.

Habitat and Ecology

The Nootka Rose prefers mostly open habitats from low to middle elevations on both sides of the Cascade Mountains. Shrubs flourish in abundant sunlight and generally drier soils. The blossoms of the Nootka Rose belong to the fullness of spring, opening with gentle breezes and warm sunshine to grace stream banks, shorelines, roadsides, thickets, and meadows.

Gesture

Of the wild Roses of Cascadia, the flowers of Nootka Rose, growing up to four inches across, are perhaps the showiest. They range from pale to vivid pink, five petals surrounding a burst of yellow stamens. As with other flowers of the *Rosa* genus, *nutkana* blossoms emanate a dewy freshness, timeless beauty, and irresistible fragrance. Like other Roses, the Nootka has strong, deep roots, and its roots and branches grow and spread vigorously. *Nutkana* is distinguished from other Roses in that it bears its flowers singularly at the ends of branched stems. Thorns are stout and grow in pairs from leaf nodes, and, overall, shrubs are larger in size, often reaching six feet tall. Five or more leaflets, always in odd numbers, alternate to form pinnately compound leaves. Leaflets are round and toothed except at the base. Graceful sepals curve into a slender tip, extending the length of the petals. Sepals join in a five-pointed star, embracing the five pink petals and sunny shower of stamens. Five-pointed stars, or pentagrams, are emblems of the flowers of the Rosaceae family, and repeat throughout their gestures.

Sepals persist after petals drop, and they extend from the bottoms of the hips, the leathery, fleshy enclosures of the dry seeds. As the red-orange hips ripen, bushes flare once again with vivid beauty from late summer into fall. The first frosts can deepen the hips into many shades of red and purple as they adhere to bare branches throughout winter and early spring. Here they offer sustenance to many creatures through the harsh times, as their nutrients, including vitamin C and flavonoids, persist through months of storms and icy cold. Seeds as well as the fleshy part of the hip are very nourishing.[1]

The tender new shoots of Nootka Rose are also a nutritious food. Once flowers are pollinated, falling petals can be gathered to sprinkle in salads or on top of cakes; they add delicate flavor and soulfulness to just about any food. Traditionally, teas made from small branches and bark strips of the Nootka Rose were used to cleanse and heal the eyes and sharpen vision. Leaves were mashed or chewed to make poultices for abscessed wounds.[2] Roses in general are valued for their astringent, cleansing, soothing, and tonifying properties. Petals and leaves have a cooling and immune enhancing effect; they offer medicine to calm inflammation, clear toxins, promote tissue repair, and subdue infections.[3]

Medicine Story

Nootka Rose is an offering of freshness and beauty. With its roots firmly anchored to the Earth, its flowers are unrestrained expressions

---

1   Elpel
2   Pojar et al.
3   McIntyre

of sensual delight. Exuberant in its growth, the thorny branches, while protecting the plant, weave a safe haven for birds and small mammals. The Rose clearly loves the Earth; as it grows it contributes to the garden of Earthly delights. Providing food, shelter, color, and fragrance, Nootka Rose thickets teem with living creatures.

The Rose imparts the wisdom of being fully present, willing to live, feel, create, and participate in life on Earth. The flowers encourage embodiment through feeding all of our senses with soft beauty. As we open and expand our sensual nature, we become more alive and connected with all life. Along with other Roses, Nootka Rose medicine uplifts the heart. Its fragrance makes us breathe deeply, and imparts well-being and comfort. The texture and color of the petals gently impact the heart, helping it expand. The *nutkana* flower is a mandala expressing harmony and beauty, the sanctity of physical form, the preciousness of life on Earth.

The Rose is a central symbol in European and Middle Eastern spiritual traditions, much as the Lotus is in the Far East. The fivefold geometry of the wild Rose is an emblem of the underlying harmony and inexhaustible regenerative capacity of life. Enfolded within the five sepals and petals of this flower is the Golden Spiral, which underlies nature's architecture. (See Chapter 3.) The Rose reflects the perfection and beauty of the human form, with our golden ratio proportions and our five overall appendages. Our hands and feet—connecting us to the world around us through creative action—also have five appendages. We have five sensory channels for partaking of the joy, pain, and pageantry of Earthly life. The Rose is the emblem of the incarnation of Divine Love in human form. Roses are associated with the embodied Christ and with the Holy Mother. They often appear in depictions of the passion of Christ, for the flowers and thorns express the opening to suffering that engenders healing and compassion.[4] Rose windows found in churches and cathedrals reflect the

---

4   Grohmann

sacred geometry and sublime beauty of the Rose, inducting worshipers into higher states of compassion and a sense of oneness with Creation.

We can enlist the help of the Nootka Rose for depression, despondency, and apathy. Its medicine strengthens our willingness to give and receive, to open to both the fulfillment and wounds of love. The gesture of the Rose teaches us to be an offering of beauty and nourishment, to be vulnerable—protected through strong embodied presence, rather than withdrawal from life. Through our bodies and emotions we are able to partake of and contribute to the feast of physical life. The beauty of the Rose instills reverence for all of existence and gratitude for the sacred temple of the body, through which we experience life on Earth.

Oregon Iris

Iris tenax

IRIDACEAE

We all move on the fringes of eternity and are sometimes granted vistas through the fabric of illusion.

__*Ansel Adams*

Botanical Description

One or 2 blue-lavender to purple flowers with yellowish throats grow from erect stems. Leaves grass-like and taller than stems. Plants 8-14 inches tall.

Range

Found in Washington west of the Cascade Mountains from southern Thurston County south through western Oregon, ending just shy of the southern Oregon border.

Habitat and Ecology

Oregon Iris grows in open habitats such as meadows, prairies, sparse oak woodlands, and logged coniferous forests. Plants thrive in sunlight to light shade, spreading in clumps as rhizomes branch. When the lowlands display their brightest colors—usually in late April to early May—travelers through the prairies south of Puget Sound are treated to roadside displays of softly radiant, lavender blossoms.

Iris tenax is more cold tolerant than other Irises, and in addition to its sea level abodes, it can grow up to 3,000 feet. In late spring to early summer, Oregon Iris appears on open flanks of foothills and mountains. In the varied elevations of the western Columbia Gorge, this Iris is abundant. Silverstar and Bluff Mountains are replete with open meadows due to the historic Yacult fire of 1902. When Avalanche Lilies begin to melt through the last snow drifts that top these mountains (roughly between summer solstice and early July), their mid elevation slopes host showy clusters of Oregon Iris.

Gesture

With its dreamlike wash of purples, striking design, and three groups of three floral parts, Oregon Iris is a testimony to Nature's artistry. As each floral group extends in different directions, the inflorescence resembles an elegant dancer. Large sepals reach out from the floral tube, then reflex downwards. Their rich color is accented by a dab of yellow that bleeds outward from the base into pure white. Small violet "style branches" arch over these sepals, bending upward just at the point where the sepal bends toward the ground. The graceful petals curve skyward.

Guided by the white and yellow "landing lights", a bee squeezes between a sepal and the downward curve of the style branch. Precisely at this curve the stigma scrapes pollen granules from the furry bee. As the insect crawls further towards the nectar at the flower's base, he is dusted with new pollen from the anther tucked between sepal and style. Once nectar has been gathered, the bee finds his way out through the arch of the style, his new pollen granules intact until an encounter with the next flower's stigma.

The species name, *Iris tenax*, refers to the toughness of the sword-shaped leaves. These leaves were valuable to the native people who shared the plant's habitat. *Tenax* comes from the same root as the word tenacious; the Latin word *tenere* means "to hold", and *tenacem* means "holding fast". The leaves, narrow and thin as well as strong, were ideal for weaving into rope and other artifacts for holding and containing. Native people fashioned them into snares strong enough to capture and hold deer and elk.[1]

In Greek lore the messenger of the goddess Hera was known as Iris, and she often appeared as the rainbow. Her name can be translated as "rainbow", also as "the eye of heaven". Iris is the name given to the colored part of the eye, the window to the soul.

The flowers of the Oregon Iris quickly develop and fade, yet the foliage is renowned for durability and strength—for tenacity. Humans have revered this plant for its practical, dependable service as well as for its uplifting, ephemeral flowers.

---

1   Mathews

Medicine Story

Like a rainbow, like a message from a goddess, the Iris flower opens our eyes to the shimmering beauty of the Earth, renewing our sense of wonder and purpose. In the plant's rapid vertical growth and sword-like leaves, it extends like a conduit between ground and sky. With its threefold patterns, the Iris flower forms a Star of Bethlehem, a symbol of the interpenetration of heaven and Earth. (See Chapter 3.) Water and crystals, mediums for light and color, display this six-pointed star in their structures. Rainbows, the brilliant crystals of gems and bird feathers, and the clear colors of flowers are messages of divine inspiration. The light-filled iris of the eye is the portal for receiving these messages.

Under the spell of an Oregon Iris blossom, a statement formed in my awareness: "The way that you see is the way that you be." Years later I learned of related words in the Talmud: "We see the world not as it is, but as we are."

Vision is complex and mysterious. Both ancient and modern seekers describe it as the interface of outer physical light with the inner light of mind and soul.[2] Our beliefs, past experiences, thoughts, moods, and surroundings affect what we are able to see; this in turn conditions our approach to the world. Habitual exposure to artificial environments erodes our awareness of the patterns of nature—patterns of harmony, connection, and wholeness. When we lose sight of the unity of creation and perceive ourselves as separate pieces in a disjointed world, responses such as fear, depression, and mistrust are understandable. These states of constriction attempt to block pain. Even our eye muscles become constricted; therefore, under the influence of a negative emotional state, the amount of light that enters our eyes is reduced. This visual distortion adversely affects every cell in our bodies as well as our relationships and sense of well-being.[3] The

---

2   Zajonc
3   Appelgren

distortion continues to reinforce itself in a negative feedback loop.

Radiant colors are nourishment for the iris of the eye.[4] They feed our souls as well, breaking through the dark cloud layers that confine our vision. Iris, whether in the form of a rainbow, the messenger of a goddess, or a flower, is the archetype of the beauty and color that successfully penetrates our dreary, constricted states. Iris, in her multi-colored guises, pierces the illusion of a disconnected, senseless world and floods us with inspiration. Her divine messages are elusive and fleeting, often just a momentary shimmer that melts with the rainbow. Yet they reorient us to the eternal truth of our souls, of a coherent world infused with meaning and color. Through restoring clear vision, Iris awakens the artist, whose expression helps others to glimpse the beauty of creation.

The medicine of Oregon Iris is illuminated vision. The flower's artistry reminds us to take time each day to drink in color with our eyes by gazing at the sky, blossoms, autumn leaves, or other pure colors that nature offers. The more brilliance we are able to absorb, the more we radiate. Iris is an emissary of what has true and enduring value, opening our eyes to a world suffused with the divine.

---

4   Kaplan

Rock Penstemon

Penstemon rupicola

SCROPHULARIACEAE

There's no way that you have enough courage, strength, and perseverance to birth and raise a child, to speak out for what matters most, or to get out of bed each morning and live your life. But the doing of these things becomes that courage, strength, perseverance, and much more.

> **Botanical Description**
>
> Mat-like plants with stems less than 4 inches tall. Leaves bluish-green, oval, finely toothed, and evergreen. One and ½ inch long tubular flowers are pink to deep rose.
>
> **Range**
>
> Grows in the Cascade Range west to the Coast Range, from central Washington through Oregon into Northern California.

Habitat and Ecology

From cliff faces, rocky outcrops, and talus, Rock Penstemon flowers cannot go unnoticed. The dark basalt of the South Cascade volcanoes is a striking backdrop for their displays. Even from afar the rock seems to ignite in vibrant patches of pink. As indicated by its common and species name (*rupicola* means "rocky"), this plant takes root in rock crevices and pockets—wherever a bit of soil might collect. Often its flowers can only be admired from a distance, their alluring color appearing along death-defying walls and ledges. Rock Penstemon grows where nutrients and water are scarce and where exposure to wind, sun, and temperature extremes is assured. Along with such challenges, these habitats afford good drainage and plenty of sun.

Rock Penstemon is considered a mountain plant, for it grows in the upper forest zone into subalpine and even alpine areas. However, in the abundant basalt cliffs and outcrops of the west side of the Columbia River Gorge, it is widely found at both low and high elevations. In early May, when Balsamroot and Lupines paint the lower hillsides of the Gorge, Rock Penstemon's blossoms glow from precipitous basalt formations. By June, it is blossoming on the higher flanks of the mountains above the Columbia. Here it coincides with a lively tapestry of flowers, including Paintbrush, Beargrass, Rosy Spirea, and Oregon Iris. Early July calls for a pilgrimage to Crater Lake, for here Rock Penstemon and Davidson's Penstemon (*P. davidsonii*) converge in habitat and bloom time. The vivid red-pink of the *rupicola* and the cooler lavender of *davidsonii* enhance each other's beauty from rocky perches above the lake's unforgettable blue. Later July and August find the *rupicola* at the higher elevations of the south Cascade volcanoes, such as Mt Rainier.

Gesture

The brilliant rose to red-pink color of the blossoms is the most prominent feature of Rock Penstemon. Long tubular flowers grow in clusters of three or four from the ends of short stems. Stems arise from misty, grayish-green mats of foliage. The flowers' vibrant color and significant size of up to one and a half inches long make them especially prominent compared to the plant's small muted leaves and short stems.

A closer look at the flowers reveals that the tubes are made up of five fused petals that flare at the ends. The flared tips, with two lobes above and three below, become the two lips of this mouth-like floral opening. Peering down the "throat" of the flower one can glimpse the relatively advanced reproductive strategy of this genus. Characteristically, Penstemons have four fertile stamens, which have anthers with sacs of pollen, and one infertile stamen, with no anther. *Penstemon* translates as "almost stamen" (*pen* = almost; *stemon* = stamen), and refers to this fifth sterile stamen, called a staminode. In the Rock Penstemon the anthers of the fertile stamens are covered with dense white wool. The anthers hang above the staminode, which begins at the roof of the floral tube and then drops to the floor near the opening. Many staminodes are hairy at the ends to glean pollen from bees, but the *rupicola*'s is smooth or with just a few golden threads. In the *rupicola* white woolly hairs line two ridges along the bottom of the floral tube. These may guide bees while scrubbing them of pollen as they proceed down the tube. *Rupicola* flowers also have a ridge on the outside top of the floral tube. Inside the tube's roof, this is a groove, in which lies the pistil, or female part. The pistil's sticky stigma, the pollen receiver, hangs over the staminode with the anthers.

The staminode may protect the ovaries of Penstemon flowers from harmful insects.[1] Its characteristics vary with different species, and its variations are responses to the pollination challenges and opportunities of the species' habitat. The staminode has co-evolved with its environment to increase pollination efficiency. Within the corolla tube it guides the path of bees, ensuring the amount of pollen that is deposited and received is the most beneficial for that particular habitat.[2] Since the stiff hairs, or "beard", which often grow at the staminode's tip, are a key part of this strategy, a hairless staminode, such as that of the *rupicola*, may indicate that, for this species, hummingbirds are more prominent pollinators than bees.[3]

These strategies demonstrate a progressed level of development. Plants with flowers with bilateral symmetry such as Penstemons are thought to have evolved from radially symmetrical (circle or wheel-shaped) ones. Five stamens fit well around the circle of a radially symmetrical flower. However, as five petals fold inward, fusing into a tube, space no longer allows for an even distribution of this many stamens. The Penstemons have resolved this issue through pairing the fertile stamens into one shorter and one longer pair, retaining an infertile fifth in the middle for its added benefit to reproduction.[4]

---

1   Strickler

2   Visalli, Dana, *Methow Naturalist*. Visalli explains that species in habitats with few pollinators need each pollinator visit to maximize the amount of pollen deposited and received. On the other hand, where pollinators are abundant, plants benefit from limiting the amount received on any given visit in order to assure that pollen from the greatest number of flowers will fertilize the ovules.

3   *Ibid.*

4   *Ibid.*

When a flower evolves from radial to bilateral symmetry, it aligns more closely with animal nature, reflecting animal shapes and employing these to create an advantageous relationship to pollinators. (See Chapter 3.) Rock Penstemon's tubular shape, with its enclosed interior, appears creature-like. Its opening is similar to a mouth with lips and a throat.

The Rock Penstemon, woody at its base, forms mats of foliage that hug the rocks. The small oval, serrated-edged leaves grow in opposite pairs. They are leathery, covered with short fine hairs, and crowded together. Their misty blue to grayish-green appearance comes from the fine blue-white waxy powder that covers their surface. Botanists refer to such waxy leaves as "glaucous". The foliage characteristics provide protection in this plant's very exposed environment. Rock Penstemons grow in areas of strong sun and wind, rapid changes and extremes of temperature, and frost heaves that disrupt the soil and roots. The leaves' fine hairs provide insulation and attract and retain water in the face of drying winds and heat. The glaucous covering protects the leaves from damage from excess sunlight. Their leathery toughness preserves them in fluctuating temperatures, enabling them to withstand intense heat and sub-freezing conditions. The leaves also protect themselves from wind and weather and preserve moisture through their small size and by crowding together. Similarly, the plant's rock-hugging growth pattern is a protective adaptation. Observation of the Rock Penstemon's foliage reveals how this plant manages to flourish in its harsh habitats.

Because of its harsh, exposed environment, Rock Penstemon grows very slowly. In some years it is able to sustain only incremental new growth. Therefore, even small plants have survived to an impressive age.

Medicine Story

Rock Penstemon's flowers and foliage noticeably contrast with each other. The flowers are large and showy. Saturated with vibrant color, petals converge in a dynamic, expressive shape. Flowers shine and capture attention even from a distance. The leaves, on the other hand, are diminutive with muted color. They huddle together and hug the ground. Without their flowers, they easily go unnoticed. Flowers are overtly expressive while leaves are protective, their traits providing shelter from environmental challenges.

Rock Penstemon lives with severe conditions, in places of rapid fluctuations and extremes. Its environment offers little in the way of shelter, soil, nutrition, and moisture. Rock Penstemon makes do in the open with a minimal deposit of soil and scant nutrients and water. Its roots take hold adjacent to bedrock. The topography here, like the weather and elements, is extreme, with sheer heights and dramatic drop offs, and plants grow from precipitous perches.

Rock Penstemon's flowers arrest us with their brilliant color. The vibrant, glowing pink speaks to the heart; it inspires and energizes. The color imbues its surroundings with a celestial radiance. However, up close, peering into the wide open mouth of Rock Penstemon's floral tubes, the flowers become expressionist art. They appear frightening, like a fierce predator, or perhaps like the terrified prey. The "mouth" seems to express intensity—fear, rage, outrage. We recognize human-animal shapes in the flower's bilateral symmetry and enfolded interior; we see a reflection of our nature, of our life dramas. The Rock Penstemon flowers seem animated; they seem to possess creature traits such as movement and emotion. Their outstretched open-mouthed gesture is dynamic, striving, assertive.

The flower essences of two species of Penstemon, Mountain Pride (*P. newberryi*) and Davidson's Penstemon (*P. davidsonii*), both relate to courage. Mountain Pride essence helps one to have the courage to act on one's convictions, to be a spiritual warrior. Davidson's Penstemon is for accessing the inner fortitude and courage to meet adversity, for tapping into previously unknown resilience and strength.[5] Both of these species have a close similarity to the Rock Penstemon, and where they come in

---

5   Kaminski, Katz

contact, Rock Penstemon hybridizes with either of these.[6] The gesture of the Rock Penstemon is one of resilience and fortitude in the face of harsh conditions. Its medicine story expresses the courage to survive, to be shaped by challenging and extreme circumstances.

In an environment that offers little support and much adversity, the foliage has responded by developing shelter and insulation from within itself. Through its interplay with its environment the plant has learned to get the most from very little, whether it be water, nutrients, or soil. The leaves are protective and conservative, while the flowers are unabashedly expressive.

In the flower's expression is another duality. Its shape and structures evoke the emotions, drama, and intensity of the animal-human experience. Yet the flower's color inspires conviction in life's beauty. Their rose-pink uplifts the spirits. One recognizes purpose and calling within hardship.

Courage is forged from many factors: adversity, support, protection, fear, fierceness, inspiration. All of these factors can propel us toward vibrant self-expression. Consider how adversity and lack of assistance in your life may be eliciting latent life-serving qualities just as Rock Penstemon's harsh habitats have shaped its evolution. Its habitats have elicited innovative and eloquent survival responses. Just as with the plant's foliage and growth patterns, our difficult life circumstances summon resilience, strength, and creativity. The dynamic beauty of the flowers invites us to hold nothing back in service to aliveness. Allow their irresistible pink to pull you to places where your sane measured self knows better than to go.

---

6   Turner, Gustafson

Upland Larkspur
Delphinium nuttallianum
RANUNCULACEAE

Many students want to "gain control" over their bodies. I have to remind them that this is not our goal. The task is to learn to listen to the intelligence of our home. When we value what the body has to tell us, we create a dialogue with our senses. The same is true for the Earth....Control limits possibilities; dialogue invites surprise.

__*Andrea Olsen*

Botanical Description

Plants 4-20 inches. Slender erect stems bear open racemes of 14 or less flowers. 5 bright blue to purple sepals; upper sepal has a straight slender spur. Petals white to pale blue or purple, occasionally yellow, with blue or purple veins; lower 2 petals are bilobed; stamens numerous. Leaves are palmately dissected with sharp tips.

Range

Grows from southern British Columbia southward to northern California, and eastward to Alberta, Montana, Wyoming, and Nebraska, south to Colorado and Arizona.

Habitat and Ecology

Upland Larkspur is a plant of dry foothills dotted with Ponderosa Pine. As it grows it requires moist soil, but it needs well-drained situations, such as slopes or gravelly areas. More prevalent east of the Cascade Range, Upland Larkspur also grows in the drier areas of the west side, such as the prairies south of Puget Sound, and the "rain shadow" islands of the northern Sound and the Strait of Juan de Fuca. One finds it in the central part of the west Columbia River Gorge, from where its range extends to the Gorge's eastern extremity.[1] This Larkspur requires ample sunshine. It flourishes in dry open woodlands, grasslands, and sagebrush steppe habitat. Growing on the light-filled slopes and meadows of mountains and foothills, it is especially abundant in the Ponderosa Pine zone.

Flowers appear with the mid-season surge of blossoms, anywhere from April to July. One may come across a few flowers sprinkled here and there, or bright blue to indigo clusters circling a Ponderosa Pine. In some areas plants paint the land with impressive swaths of color. Following one very wet winter and spring, the rolling slopes of the Blue Mountains were

---

1   Jolley

indeed blue! This species casts beautiful color over the landscape through its ability to spread aggressively, a quality for which it has been labeled invasive. It curtails the growth of other plants, notably legumes.[2]

Upland Larkspur and other Larkspurs, because of their toxicity, have been the target of human eradication efforts, including the use of soil sterilants. Larkspurs are thought to be the greatest cause of cattle death on national forest land. Though a cow must consume three percent of its body weight for a lethal dose of these plants, reports tell of over a hundred animals killed in but a few days. Cattle seem to favor Larkspurs and seek them out. Elk, on the other hand, avoid these plants early in their season, while freely consuming them during late summer and fall.[3] This feeding cycle protects the elk, for the plants' alkaloids are most prevalent during rapid spring growth and leaf formation, while after blossoming, they lose their toxicity.[4] Larkspurs are both appealing and nontoxic to sheep, so sheep herds have been enlisted to clear them from range lands.[5]

Humans have also employed Larkspur's toxicity for benefit. For thousands of years seeds of this genus have been pulverized for use as a poison for lice.[6] Delphinine, whose name is derived from the genus name, is the main toxic alkaloid among several others produced by these plants.[7] Extracts of delphinine or tinctures of the whole plant are used for ridding external areas of parasites. However, use on irritated or broken skin is contraindicated in order to avoid the absorption of dangerous toxins into the body.[8] Delpinine and the other poisonous alkaloids found in all parts

---

2   www.ibiblio.org/pfaf/cgi-bin/arr_html?Delphinium+nuttallianum
3   Craighead et al.
4   Elpel
5   Clark; Craighead et al.
6   Mathews
7   Taylor, Douglas
8   Craighead et al.; Elpel

of Larkspur plants are known as terpenoids, and their danger is in their depressive effect on the central nervous system. Trained professionals have judiciously used these substances as antispasmodics and sedatives, for they sedate the respiratory system, the heart, and the nervous system.[9] One species of *Delphinium, staphisagria,* is formulated into a major homeopathic remedy. While the physical properties of *Delphinium* suppress and sedate, this species' homeopathic addresses problems that arise from suppression, especially of intense feelings. Staphisagria is helpful for mental, emotional, and physical conditions resulting from inhibited emotional expression. People who need this remedy are sensitive; they hold back their feelings so as not to hurt others, even when they have been abused. Acutely sensitive to violation, they experience a powerless anger.[10]

Humans have also used the lovely flowers of Larkspur to make a blue dye and ink.[11] The flowers attract both bumblebees and hummingbirds as their pollinators.[12] More than once, coming upon Larkspurs has also meant an opportunity to observe nectar-gathering sphinx moths, unperturbed, apparently magnetized by indigo-blue.

Gesture

Both the beautiful color and fanciful shape of the Larkspur flower draw people and pollinators for a closer encounter. The vivid blue, rich purple, or indigo color comes from the five sepals. Larger than the petals, the sepals spread open, their distribution resembling human appendages. The top sepal has a slender nectary spur extending behind like the flourish of a jester's cap. Four petals, comprised of two unlike pairs, grow in the heart of the flower. Smaller and lighter in color than the

---

9   Elpel
10 Curtis, Fraser
11 Elpel
12 cat.inist.fr/?aModele=afficheN&cpsidt=1077467

sepals, the petals are less conspicuous. The upper two have nectar spurs that grow within the sepal spur. In *Delphinium nuttallianum* the petals are white, sometimes yellowish, or pale blue to pale purple. These petals are inscribed with blue or purple veins, and the lower two are prominently divided into two lobes. The spur of the Upland Larkspur is straight and slender, and the sepals are reflexed, which gives them a wide-open expressive gesture. All the flower parts of Larkspurs and other Ranunculaceae are independently attached rather than fused, a trait which connects them to the most ancient flowers as evidenced in fossil records.[13]

Upland Larkspur's stem is strongly upright, bearing its flowers in open racemes. A few leaves grow from long petioles mainly from the lower part of the stem. Deeply divided into lobes with sharp tips, they resemble outstretched hands, fingers extended. Leaves develop early and the lowest ones often wither by the time the flowers open.

Because of its widespread prevalence, Upland Larkspur is also called Common Larkspur. Other names, Bilobed Delphinium or Two-lobe Larkspur, describe the lower petals' double lobes, a key identification trait. The genus name, *Delphinium*, is attributed to first century A.D. Greek physician, Dioscorides. It is derived from *delphin*, the Greek word for dolphin.[14] The flower buds resemble a swimming creature, for the nectary spur extends behind like a tail. The buds are a blocky oval shape, much like the body of a cetacean. When open, the sepals extend like human arms, legs, and head; with the spur for a cap, they become dancing clowns.

Medicine Story

Along with other *Delphiniums*, Upland Larkspur's flower enters our imagination as an animated creature. Could it be swimming, dancing, clowning...? Its bilateral symmetry mimics animal forms; its five spreading

---

13 Elpel
14 Clark

sepals reflect the structures of humans and other mammals. Its leaves, like open palms, continue this gesture. Upland Larkspur's assertive growth habits, its ability to sweep through tracts of land, is also animal-like. Its toxic constituents dramatically impact animal-human functions, targeting the central nervous system. This system is at the heart of our animation, an interface of our physical, mental, and emotional aspects. Humans have sought Larkspur's physical and homeopathic actions for sedation and the consequences of sedation, respectively. Larkspur has elicited intense responses from humans, both herbicidal hatred for poisoned livestock and awe and gratitude for its breathtaking expanses of color. Its aggression and exclusion of other plants arouses the same traits in humans, determined to achieve its eradication. Or we employ its lethal properties in our territorial battles with parasites.

The Flower Essence Society makes a flower essence from *Delphinium nuttallianum*. It targets the imbalances of leadership, assisting those burdened by duty and responsibility or self-serving leaders inflated with grandiosity. This essence helps one enlist charisma and leadership ability to inspire, motivate, and encourage others, in service of worthy causes and ideals.[15]

Because of our ability to move, grasp, and communicate, humans are capable of dramatically impacting our surroundings. We can also experience a colorful palette of emotions. The many facets of Larkspur's gesture mirror our excesses and suppressions. Through this plant's homeopathic we find the relationship between sedation and forceful outbursts. These outbursts take the form of symptoms and disturbances on personal and collective levels. Larkspur's medicine helps us grasp the enormity of the power we wield through the gifts of motion and emotion, thought and action, the ability to feel and respond. Ungraced with the instinct of the elk, we are able to function outside natural checks and balances.

---

15 Kaminski, Katz

This freedom augments our power, our capacity for great service or great harm. With trial and error as our teachers, we are awakening to the dire consequences of domination. We attempt to master our bodies, to bend the Earth to our will, rather than discerning the wisdom of natural systems, allowing them to lead us. Perhaps we can take our cue from the dolphin (and dancer) the Larkspur bud and flower show to our imagination. Rather than seek dominion, dolphins apply freedom and intelligence to cooperation, play, and sensual pleasures. As we discover right relationship to power, we begin to experience the true potential of our delicate and complex systems.

Simpson's Hedgehog Cactus
Pediocactus simpsonii var. *robustior*
CACTACEAE

Growth occurs through gathering, absorbing, consuming, and accumulating—also through opening. Then, light, air, and emptiness come pouring in.

> **Botanical Description**
>
> Bulbous stems, single or in clumps, have radiating clusters of spines. Flowers vivid rose to magenta with numerous bright yellow stamens and 1 pistil. Flowers grow from the tops of stems.
>
> **Range**
>
> Found in very dry areas of eastern Washington, eastern Oregon, north central Idaho, and north central Nevada.

Habitat and Ecology

On the driest of the basalt-strewn hillsides and ridges of eastern Washington known as lithosol—or rock soil—grows the state's only Hedgehog (or Ball) Cactus, the *robustior* variety of the Simpson's Hedgehog. This spiny plant at first blends with the brittle, mottled groundscape, but once its round fullness comes into focus it begins to stand out amongst sparse, contracted plants. One might wonder how anything could grow in such a habitat, short on soil and water, severe with sun and rocks.

On closer observation it becomes clear that the life that flourishes here does so because of the rocks. The two areas where this Hedgehog is mainly found, the Beezley Hills and the eastern portions of Umtanum Ridge above the Yakima River canyon, receive an average of only seven or eight inches of rain per year. The living communities here are shaped by the intrinsic qualities of water. When rainfall occurs, the basalt rubble becomes a water catchment system for the slopes and ridges. Rocky dips and pockets become cradles for life. Plants here grow in clumps, and they too become part of the catchment strategy. The rocks and plant clumps reflect on a smaller scale the patterns of the overall landscape, where the hills' receptive folds become green with spring vegetation.

The fleshy body of the Hedgehog is its own water catchment system. Its essence is moisture containment, and it holds the story of its landscape's relationship to water. Accumulation and protection are its survival strategy, and its moist interior is infrequently pillaged by an adept critter. Its lovely flowers, however, are offered freely to many, including green pollinating flies and tiny beetles who circulate among the stamens.

Robustior flowers open during the peak of the spring bloom, which occurs in late April in the eastern and lower areas of its range and in mid May in its more western and higher abodes. Also blooming near this Hedgehog in the sparse rocky areas are Thyme Buckwheat, Bitterroot, and Gairdner's Penstemon. The folds of the hills, where more moisture and soils accumulate, host luxuriant stands of Arrowleaf Balsamroot and Large-Leaved Lupine. The air is bright with the liquid songs of horned larks and meadowlarks.

Gesture

Round and impervious as the hills on which they grow, these sage-colored cacti are a response to a habitat where water is scarce. As though its stem needed to be more container than conduit, a globular form replaces a streamlined channel. What would be branches or petioles become bumps called tubercles, and in place of the opening of leaves is the precision of needles. The plant creates a reservoir in its center as it parcels the flow of moisture and life energy to its extremities. The needles contrast with the mass of the body. Minimal in form, needles grow in clusters and radiate outward from each tubercle. Arranged in spirals, these clusters form an open yet impenetrable network over the sphere of the cactus. Sunlight plays among the needles, and at the edges of day the cactus glows with an orange-gold aura.

Pink buds develop in a spiral pattern from the top of the cactus sphere. Colorful red spines herald their arrival, surrounding the places from which they emerge. At first buds resemble tubercles, but they soon elongate, unbound by centripetal, contracting forces. Like the needle clusters, Hedgehog flowers radiate open, reaching upwards to air and sunlight. The *robustior* does not restrain itself in its flowers; it is as though its moisture and life force have been hoarded just for these magenta petals and glowing stamens. The mass of yellow stamens are a splash of light in the center of waxy pink petals, a showy contrast beckoning visitors. The flowers open in response to the sun, closing at day's end or when a cloud passes. In this way they, too, express the conservative, protective nature of the plant. Blooming time is short, and in a matter of weeks knarly, elongated seeds have replaced withered flowers. They are the size and shape of the buds, but light brown in color and very dense. Concentrated with life and nutrients, seeds await opportunities to break open.

When one comes across the rare Hedgehog that has been bitten or somehow damaged, there is opportunity to view and feel its soft green interior. Spongy, moist, and gelatinous, it is a cooling balm. No wonder a hard green shell and fortress of spines protect such a delicacy!

Medicine Story

A plant species reveals its unique essence—and therefore medicine—through the ways in which it improvises on or departs from the characteristic gestures of its genus or family, or of plants in general. All species of the Cactus family are extreme variations of the archetypal plant. Cactaceae are well adapted to arid environments. Accumulation and protection are at the heart of their survival strategy. Fleshy, hard-shelled growth replaces stems, and instead of leaves unfolding, harsh needles protrude. The gesture of accumulation of water is very pronounced in the Hedgehog Cactus. These plants, as demonstrated by the Simpson's Hedgehog, tend to produce very bulbous forms. Hedgehog flowers are often striking in color and form, as the *robustior's* showy flowers demonstrate.

Simpson's Hedgehog expresses two kinds of expansion. Its spherical trunk is its dominant form. Its watery and earthy nature is revealed in its roundness, in its capacity to absorb and hold quantities of water and nutrients, to amass substance. This cactus assumes the look of the small basalt boulders that it grows among, and echoes the shape of the rolling hills. It holds a piece of the story of the great sphere of the Earth, whose curve is visible in this expansive landscape where *robustior* grows.

The second way of expanding is found in the spines and flowers of the Hedgehog. Instead of gathering in, they radiate outwards, opening to light and air. In this gesture the sun-drenched and windswept qualities of its open habitat permeate the cactus. The Hedgehog moves in its unique way between holding and releasing, accumulation and sparseness, vulnerability and protection. These poles live in a rhythmic connection within this plant. The trunk is bulbous, the needles are sparse. The plant amasses moisture and nutrients, and its vulnerable flesh requires spiny protection. Like its landscape, the *robustior* is both harshness and beauty.

This cactus invites us to reflect on our own patterns of protection and vulnerability, of holding and releasing. How do we respond to scarcity? Where in our bodies and in what ways do we hold, and where and how do we let go? Can we fully receive and absorb nutrition? How does accumulation make us vulnerable, and when is it most beneficial to release or give away? When do we grow through gathering and acquisition and when through opening and emptying? We absorb water, nutrition, and love; while we open our minds, let go of preconceptions, and expand our horizons. Meditating on the unique adaptation and unusual beauty of this cactus helps you work consciously and find your own balance with these rhythms.

Large-leaved Lupine
Lupinus polyphyllus
FABACEAE

The cosmic mind whispers to us in the silent spaces between our thoughts and there is a sudden knowingness and we are transformed.

—Deepak Chopra

Botanical Description

A perennial herb with erect stems up to 4 feet tall. Basal leaves are longer than the stem leaves creating a canopy of leaves under the floral raceme. Leaflets are pointed at the tip, numbering 9-17. They are smooth on top and hairy below. The flowers form a dense, stalked terminal raceme up to 1.5 feet tall. The raceme has whorled flowers near the top and scattered flowers below. Corollas are bluish to violet and smooth.

Range

Grows from British Columbia into California, from the coast into the Cascade Mountains and eastward to the shrub-steppe.

Habitat and Ecology

The Large-leaved Lupine adapts to many conditions; therefore, it can be found in a variety of habitats. It grows in the open or in shade, in wet or dry areas, from lowlands to high mountain meadows. Many flower guides describe this plant's affinity for wetness, its prevalence in moist meadows, by stream sides, and even in bogs. Yet it is also able to flourish in dry habitats such as shrub-steppe. Through special characteristics, the Large-leaved Lupine is able to fulfill its requirement for ample moisture even in some of the driest terrain.

One can observe this adaptation in the Beezley Hills, a lithosol area in eastern Washington. Lithosol—or rock soil—habitats are among the driest of the Pacific Northwest. Large-leaved Lupine grows in abundance in the Beezley Hills, in May painting the land with indigo. It settles into the green folds of the hills, where its purple spires contrast with the bright yellow pinwheels of Arrowleaf Balsamroot. Following the paths of snow melt and spring rain runoff it knows how to get the most from these seasonal streams and seeps. To begin with, its leaf pedicels vary in length such that all the leaves meet at the same level, forming one dense layer, or canopy, just below the flower stalks. In addition, each leaflet is smooth on top while it has fine hairs underneath. This distribution of hairs is a distinguishing feature of *Lupinus polyphyllus*. This Lupine also has more leaflets per compound leaf than other Lupines. Leaflets radiate into a wide circle, resembling a great hand whose palms remain slightly cupped. With their smooth, waxed paper-like tops, the leaflets' design is ideal for capturing rain in the cupped center, shaping it with their contours into a floret of water. As the hot hours of afternoon sunshine bake the land, this water remains. Meanwhile the leaf canopy is shading and preserving the ground moisture below. The soft hairs on the undersides of the leaflets capture droplets of condensation in this greenhouse-like zone. From above and below the large leaves are perfectly suited to secure and preserve water, forming a kind of green river. As they enhance their own survival, they contribute to the water catchment and preservation system of their habitat.

In addition, like all members of the Fabaceae family, Large-leaved Lupine hosts bacteria in its root nodules. These bacteria draw nitrogen from the air and transmit it to the soil in a form that plants can assimilate. The entire habitat is enriched as soil improves and nitrogen and protein are increased in the plants' tissue. Lupines provide important forage for chipmunks, deer, and squirrels, however, plants contain alkaloids toxic to humans.

Gesture

The adjective "stately" is often applied to the distinctive spires of blue to violet flowers of the Large-leaved Lupine. These upright racemes tower above those of other Lupines. Showy blossoms grow up the spike, first in spirals, then whorls repeat the radial pattern of the leaves. No wonder most Lupine cultivars are developed from this species.

The five-petaled Lupine blossom is an ingenious pollination strategy. A broad upper petal, known as the banner, is flanked by two smaller "wings". The banner is blue with a white to pale yellow patch. Indigo spots appear just at the base of this light area where the banner joins the keel, the two fused bottom petals. The contrasting spots guide the bee's landing to just where his weight on the keel releases the curved sheath of stamens and pistil. The sheath strikes the bee's abdomen and pollen is exchanged between fur, stigma, and anthers. After pollination the banner's white or yellow patch turns to red violet. The Lupine flower has evolved as a relationship with the bee. The bee's animal nature is reflected in the flower's bilateral symmetry, enclosed interior space, color change, and movable parts.

Medicine Story

The Large-leaved Lupine is a conduit between air and soil, sky and Earth. Its leaves open into receptive palms, resembling satellite antennae, and indigo spires commune with the rarefied air. It captures both water—runoff that otherwise would quickly dissipate—and nitrogen, and conveys them to its greater environment in usable, nourishing forms. It is a receiver and transformer, like the archetype of the water bearer, who gathers the water of the stars and bestows it on the land. In this archetype the water is imprinted with the evolutionary codes from the far corners of the galaxy, so the water bearer helps to seed the Earth with radical wisdom. Water is the quintessential conduit between above and below, sky and Earth, as evident in its hydrological cycle. This movement between etheric and physical is mirrored in water's energetic properties, for water records, stores, and transmits informational patterns to living tissue.[1] Various constellations of the sun, moon, and planets, as well as the patterns of vibrant, healthy habitats, imprint water with codes that assist growth and well-being.[2] The jeweled spheres left by rain and dew in the palms of its leaves beautifully express Lupine's gifts.

Large-leaved Lupine's impressive spires and receptive leaves express its ability to gather something rarefied, just beyond reach, and to instill it

---

1 Greene; Ryrie
2 Schwenck

in an absorbable, beneficial form. Its medicine evokes the role of artists, prophets, and visionary leaders.

Lupine's gesture and contributions participate in a deep ecology. The plant, its pollinators, and the surrounding land are a harmonious system, which simultaneously enriches each participant and the whole environment. Awareness and appreciation of each interwoven community enhance our own well being, and we convey its life-giving rhythms to the greater world.

Sagebrush Mariposa

Calochortus macrocarpus

LILIACEAE

If you want to study water, you do not go to the Amazon or to Seattle. You come here, to the driest land.

—Craig Childs

> **Botanical Description**
>
> A large perennial tulip-like lily. Sepals are long and narrow. Petals are pink to deep lavender, tapering to a point. Petals have a median green band and yellowish semi-oval gland inside each base. Plant grows to 24 inches tall. Basal leaf is very narrow and begins to wither when blooming.
>
> **Range**
>
> Grows in dry areas from southern British Columbia to northeastern California, extending east to Idaho and southwestern Montana. It also grows in northern Mexico.

Habitat and Ecology

The Sagebrush Mariposa blooms in the shrub-steppe regions of eastern Washington after the crescendo of spring color has been baked to brittle brown. The flowers glow like small tulips from barren slopes, in the hardpan banks of road cuts, and among Sagebrush and the ripening seed pods of withered Lupines. During this time, as spring waxes into summer, these beautiful flowers also appear on dry hillsides of the eastern Columbia Gorge, in the open or amongst sparse oak woodlands.

Gesture

Of all the species of *Calochortus* native to Washington, the *macrocarpus* bears the largest, most eye-catching flowers. The three rounded petals characteristic of this genus are colored rose to lavender, and in the center of each is a stripe of soft green. Enfolded by three longer, pointed sepals, the petals form a graceful cup. Opening wider, they become an ethereal butterfly perched to take flight. (Mariposa, the common name for this genus, is Spanish for butterfly.) The flowers are held aloft on tall stems, ensuring the elegant blossoms are not to be missed. This stately plant flourishes in intense heat, even in soil scraped by the bulldozer's blade.

Sagebrush Mariposa blooms when and where the fire element is strong, yet its gesture speaks of water. A member of the Lily family, the plant is nourished by its bulb, a fleshy womb-like structure, rich with life-sustaining fluids. This bulb enables survival where water is scarce. The cupped flowers also suggest the capacity to contain and offer water and nourishment. The six stamens and six combined petals and sepals are part of the interplay of hexagonal patterns found throughout the design of the flower. Water molecules, because of their structure, hook together in hexagonal forms, hence the six-fold patterns of ice crystals and snowflakes.[1] Plants whose floral parts appear in threes and sixes hint at their kinship with water. (See Chapter 3.)

The crescent moon pattern formed by the gland on the lower inside of each petal continues the themes of receptivity, reflection, and nourishment. The crescent patterns take part in the intricate mandala of soft yellow, ice-blue, and mauve inside the chalice of the flower. It is as though the flower offers an elixir to drink with the eyes. One could spend the better part of a day sampling the variations of design and color of each floral cup. Many showy beetles and other insects gather within, and, after precious summer showers, little pools of water collect.

In late June of 2007 Sagebrush Mariposas treated us to the most abundant displays we have ever experienced. Their soft blossoms appeared throughout the shrub-steppe meadows of Okanogan County, in some places clustering over wide areas. We found single plants with five and six blooms. Petal color was unusually variable, and the desert was graced with rose pink, iris purple, peach, even white. Inside each chalice angels, stars, and colors spun; from one to the next the kaleidescope re-configured in a multitude of patterns.

The fresh beauty of the Sagebrush Mariposa is dramatic and welcome in its scorched landscape. The artistry of the flower reflects the mastery,

---

1   Ryrie

the adaptive achievement, of this species. This plant knows how to flourish in heat, dryness, and damaged and depleted soils. It is able to blossom in circumstances that otherwise create a wasteland. The Sagebrush Mariposa's ability to store water enables it to befriend harsh elements. Because it can grow where most plants cannot take hold, and its blossoms open after other foliage has gone to seed, it sustains pollinators and brings vibrancy to sparse, scorched areas.

Medicine Story

Sagebrush Mariposa's habitat and showy flowers express the fire element. Fire relates to creative expression, inspiration, magnetism, ascension, and transformation. A plant expresses light and fire in its flower. The Sagebrush Mariposa flowers are dominant in the plant's gesture. They incorporate and celebrate the summer sunlight in the rosy glow of their chalices. Within this plant is the synergy of fire and water; it uplifts and inspires while offering a soulful, soothing quality.

The anthers, ovary, stigma, glands, and petal coloration of the Sagebrush Mariposa interface in hexagonal designs. When visiting this flower one is drawn into the beauty of these patterns; with heightened awareness, one connects to the creative power of nature. Entrained by its shapes and colors, one might wonder how vision influences the endocrine system's hormonal communications with physical functions. What roles do color and geometry play in these messages?

The cupped flower with its six-fold patterns and the plant's fleshy bulb evoke the water element. They echo the themes of receptivity, nourishment, containment, and life support. These themes are also present in this plant's ability to flourish in its hot, dry environment. Its challenges and creative resolutions offer wisdom for our human journey. Its adaptive story and each aspect of the plant express qualities associated with the feminine archetype, with the creative power to bear and nourish life. This plant informs our capacity to receive, reflect, nurture and be nurtured, and to provide emotional containment and a refuge of harmony.

The Sagebrush Mariposa is an offering of beauty in heat-scorched, often damaged places. Its lovely chalices reflect the Holy Mother, whose presence is found in the harshest conditions by those who suffer. Within desolation and bereavement is her abode. As we sit with this plant, we may ask for assistance in offering beauty and compassion in times of despair, in providing nourishment for ourselves and others. The way of Sagebrush Mariposa guides us in sustaining vision and compassion when this is most challenging and most necessary.

Great Blazing Star
Mentzelia laevicaulis
LOASACEAE

Think of yourself as an incandescent power, illuminated and perhaps forever talked to by God and his messengers.
— *Brenda Ueland*

> **Botanical Description**
>
> A shrubby plant growing to 36 inches with whitish, branching stems and dandelion-like leaves. Leaves and stems are covered with coarse, barbed hairs. Terminal flowers have 5 petals, 5 sepals, and many stamens of different lengths.
>
> **Range**
>
> Grows from British Columbia south through the eastern Cascades into California, east to Montana and south through Utah and Wyoming.

Habitat and Ecology

The Great Blazing Star belongs to arid, heat-scorched wastelands, including sandy washes, gravel bars, road banks, and sagebrush steppe areas. It grows in valleys and foothills, and in the Sierras it can be found in dry conifer forests up to 9000 feet. Just when the land seems no longer hospitable to color and foliage, the Blazing Star bursts into flower. As summer heat is mounting and the peak bloom has turned to ripening seed heads, travelers can be surprised by a dazzling roadside display. The Sagebrush Mariposa also blooms at this time, and may be growing from the same hardpan soil. Both plants have large and captivating flowers. Both are at home in hot, dry, barren places, even in the swath left by the bulldozer's blade.

Gesture

Everything about this flower is big, bright, and radiating. Five lance-like petals of satiny yellow extend wide open; from their center bursts a cluster of stamens, whose long lemony filaments are a shower of light. At three to four inches across and growing from the top of the plant and branch tips, these desert star dancers are not to be missed.

By contrast, the foliage is rough and muted. Dandelion-like leaves are deeply lobed and course to the touch. Tiny barbed hairs, which cover the stems as well, cause them to cling on contact. Plants can grow three or four feet tall, and their stout, silvery stems tend to branch and spread, even grow woody. The solid structure of the branches and the plant's ability to expand in height and breadth, seem to defy its arid, wasteland habitat.

Medicine Story

The Great Blazing Star possesses both the power and mass of the earth element and the ethereal quality of pure light. The plant reflects the course earthiness of its desert environments. It partakes of the wisdom of the desert. Schooled by sparseness and intense conditions, it has learned not only to survive but to grow large and sturdy. This plant's rugged ways and undistinguished foliage do not diminish the transcendent glory of its striking flowers. Like the desert sun in midsummer, the flowers of the Great Blazing Star hold nothing back.

During a trip to southern California, passing through the foothills east of Sacramento, I noticed the once rugged land carved into tidy ranchettes. I began to feel drained and despondent. All at once on the highway's gravelly banks I spotted a profusion of Blazing Star plants in full flower. I stopped to sit among their starry radiance and immediately my spirits lifted. Grateful to this plant who flowers even here, adjacent to asphalt and barbed wire, I am again willing to participate more fully in this moment.

The Great Blazing Star helps us reconnect with the radiance of our visions and spirits. In a wasteland, during the time when the sun scorches the land, its flowers are opening, celebrating the beauty of heat and light. It teaches us to hold nothing back in our creative expression, even in inhospitable circumstances. Blazing Star harmonizes the forces of earth and fire. The practical requirements of survival and growth are met in service to light and beauty; transcendent vision is grounded into creative form. This plant takes root in nutrient-poor, hard-packed land, and its flowers court the blazing sun. It gathers the course and ethereal qualities from its environment, embodying both as beauty.

Mountain Lady's Slipper
Cypripedium montanum
ORCHIDACEAE

Now I become myself.
It's taken time, many years and places.
I have been dissolved and shaken,
Worn other people's faces....

—*May Sarton*

> **Botanical Description**
>
> One to 3 flowers grow from an erect stem. Three sepals and 2 petals are twisted and brownish-purple; lip is white, 1inch or more long. Parallel-veined leaves are alternate, clasping the stem.
>
> **Range**
>
> Grows in the mountains of California, from Monterey and Yosemite north to British Columbia and southern Alaska. Found as far east as Alberta, south through Wyoming.

Habitat and Ecology

One may consult wildflower guides to help locate the Mountain Lady's Slipper. One learns that this plant is more likely to be found in the dry open forests east of the Cascades, but that it also grows in the denser, moister west side forests. In both cases it is to be found among mixed conifers, with some deciduous elements. One reads that the eastern and western Columbia Gorge are home to this plant, and that it favors rich humus. It will grow in disturbed areas, such as along roadsides, if enough humus is present. Guidebooks explain that though it mainly grows at lower and middle elevations, its species' name associates it with mountain slopes, for it sometimes makes its way into subalpine areas.

With these descriptions one may now undertake an informed quest for Mountain Lady's Slipper. And so begin many years of hiking and searching, during which time one is very fortunate to have a single encounter with this enchanting plant.

The reason for this elusiveness—Mountain Lady's Slipper is indeed a beautiful, enchanting plant, and human actions have entered into its habitat and ecology equation. Humans have substantially depleted this regional treasure through cutting flowers and digging up plants. This behavior spells trouble for any plant species, but it is devastating for an

Orchid. Like other Orchids, the Mountain Lady's Slipper requires soil fungi for its survival and growth. (See also **Calypso Orchid**.) Its seeds, themselves lacking in stored nutrients, need to be penetrated by a certain species of fungus, which then provides food for each germinating seed. In addition, fungal hormones suppress root hairs, instead stimulating mycorrhizal development. Mycorrhiza (combined fungal-root structures) are extremely efficient nutrient and moisture absorbers, and they connect the Orchid and the fungus to the benefits of a whole network of life forms.[1] Therefore, Mountain Lady's Slippers, along with other Orchids, can only grow in their correct soil environments. To maximize the opportunities for seeds to disperse into viable environments, Orchids produce enormous quantities of seed. Each Orchid's intricate strategy to maximize pollination and seed production involves one or very few pollinators who are a custom fit. Species of the *Cypripedium* genus are mainly linked to a few species of bees.[2] Since Orchids are habitat and pollination specialists, they require very particular circumstances in order to propagate and survive. Since Mountain Lady's Slipper is one of the largest, showiest Orchids of the Pacific Northwest, human behavior has undermined its survival needs, decimating its populations.

Gesture

The flower of Mountain Lady's Slipper is characteristic of Orchidaceae in that it has three sepals and three petals, the middle of which is distinct, forming a "lip". In the genus *Cypripedium*, this lip is inflated, shaped like a bulbous, fanciful slipper, and the lower two sepals are at least partially fused underneath it. In the *montanum*, this "slipper" petal is glowing white, often tinged with purplish veins. The sepals and other petals are bronze to maroon. Long and slender, they twist in lovely spirals, which can only completely charm the

---

1 Mathews
2 Stokes

observer. The genus name, *Cypripedium*, means the slipper of Aphrodite, or Venus, the goddess of love, beauty, and harmony. Since she was believed to have been born on the island of Cypress, the Greek word, *Kypris* was another name for this evocative goddess. *Podion* means a small foot or slipper.

Another characteristic of *Cypripedium* is its three stamens, two fertile and one infertile, the latter arching over the other two, forming a shiny structure at the base of the slipper. In the *montanum*, this structure's bright contrasting yellow may well direct a bee to the fertile stamens, which, as with other Orchids, are fused with the pistil into a single column attached to the inferior ovary.

Along with its imaginative forms and colors, Mountain Lady's Slipper's flower carries a delightful scent, insuring its irresistibility to humans and pollinators. The flower's slipper is a tantalizing "funhouse tunnel for bees".[3] The ingenious design of the flower ensures that the gratification of the bees' desires is a hand in glove (bee in slipper) fit with the plant's reproductive needs. This partnership offers a revelation for humans. In nature, desires serve greater systems and are gratified in mutually beneficial relationships.

Unlike many Orchids, the Mountain Lady's Slipper has vibrant green foliage. Lush, elongated oval leaves have prominent parallel veins. They clasp the erect stem as they spiral upwards. This Orchid, therefore, reaps the benefits of its own photosynthesis. The hairless roots depend on the resulting carbohydrates, a crucial source of nourishment which is abruptly cut off if one picks the leafy flower stems.

It can take the Mountain Lady's Slipper about fifteen years to produce a single flower. The plants with two or three flowers are decades old. A plant that reaches its maximum height of around two feet, is over thirty years old, a rare treasure.[4]

---

3   *Ibid.* p. 220

4   Rathbun-Holstein

As a genus, *Cypripedium* has played a role in herbal medicine, its various species having similar actions and applications. Since human harvesting has already put this genus in jeopardy, only the cultivated plants should be used. Michael Moore recommends the Stream Orchid (*Epipactis gigantea*) as the appropriate choice for the same, though "feebler", results.[5] *Cypripedium's* role in natural medicine uncovers a deeper picture of its gesture.

Lady's Slipper's herbal action targets the emotions, nervous system, and general sense of well-being. This plant has been used to uplift moods, sooth the nerves, and ease pain. It assists physical symptoms associated with stress and emotional upset and tension. As an antispasmotic, it relaxes cramping pains, bringing relief for tense muscles and during childbirth, menstruation, and digestive disturbance. It relieves tension headaches, palpitations, insomnia, and exhaustion. Lady's Slipper is a nervine, relieving anxiety and supporting an overwhelmed nervous system. Herbalists have used it to help people out of despair and dark depression. It is helpful when emotions catapult one into a downward spiral, and when one is hypersensitive with no tolerance for sensory input.[6] Its homeopathic action is similar to that of the herbal tincture. A homeopathic preparation from the root calms the nerves and soothes insomnia, racing thoughts, restlessness, sexual depletion, and premenstrual tensions.[7]

As a flower essence, Lady's Slipper is used to help people ground and integrate spiritual guidance into their bodies, putting it into action; it helps them actualize their authentic purpose in daily life. The essence prompts one to claim inner authority and to walk one's true path. In this way it helps address an underlying cause of restlessness, debility, and anxiety.[8]

---

5   Moore, p. 235
6   *Ibid.*
7   McIntyre
8   Kaminski, Katz

Medicine Story

Another name for the Mountain Lady's Slipper is Moccasin Flower. This flower conjures something we can slip onto our feet and walk in. According to the Doctrine of Signatures,[9] a plant whose flower resembles footwear might have some effect on the feet, whether physically or metaphorically. The plant's herbal and flower essence applications help to calm, center, and revitalize, assisting one to stand in one's own authority, to follow one's authentic path. When we are depressed, restless, anxious, and depleted, we are in need of both grounding and inspiration. When these conditions are chronic, we need to reconnect with our higher purpose for life and take the steps necessary to put it into action.

Mountain Lady's Slipper has a sublime, transcendent quality. Each stem supports a spiral stairway of green leaves culminating in a luminous slipper. At the same time, its underground mycorrhizal network expands horizontally, joining the plant with other life forms. Its flower, irresistible to humans and insects, mimics the animal-human world through its bilateral symmetry and enclosed interior space. Its shape plays with our imaginations and connects us to aspects of ourselves, to stories, movement, and emotions. The flower becomes a slipper, a path, a next step, a place to stand, a correct fit, a journey. This sublime flower grounds us into our feet, where we take action on spiritual guidance, where we move forward according to our own truth. Through our upright spines we receive spiritual transmissions; through our hands and feet we reach and move outward, connecting with others and manifesting our purpose in the world. The Mountain Lady's Slipper expresses both of these orientations.

---

9   The Doctrine of Signatures was formulated by Paracelsus, sixteenth century physician and alchemist. It asserts that outer physical forms correspond to inner qualities. Therefore, in traditional European herbalism, one examines the shapes and designs of a plant to foster understanding of its medicinal properties and how it might assist humans.

Collectively we are beginning to realize that we have stumbled off the path. We find ourselves stomping over and crushing the remains of what we love. Nature has freely offered her bounty, yet all that has shared her strongest medicine and most eloquent beauty, we have most drastically plundered. Whatever has especially captivated and called to us, we have extracted and seized in order to possess. The spirits of the disappeared and disappearing have themselves become guideposts for awakening our hearts and reorienting our steps. Our loss is at times unbearable. Now more than ever we begin to grasp each irreplaceable mystery. May the light of this clarity illumine the way back.

Purple Virgin's Bower
Clematis columbiana
RANUNCULACEAE

We are here not only to transform the world but also to be transformed.

—Parker J. Palmer

> **Botanical Description**
>
> Semi-woody vine. Flowers have 4 lavender-blue sepals and no petals, and are 2-3 inches across. Leaves are compound and opposite.
>
> **Range**
>
> Grows from British Columbia into Oregon, from the eastern Cascades into the Rocky Mountains. Found from valleys up to 8500 feet.

Habitat and Ecology

Purple Virgin's Bower has a wide range, covering much of the northern part of the intermountain West, spanning mid to high elevations. However, it is not an abundant plant, and its discovery in moist woodlands, among thickets and tree branches, or in rocky subalpine habitats, is a cherished occurrence. In April one might find its amethyst flowers woven among the spring green of a brushy side canyon of Idaho's Clearwater River. By late June its viney mats drape over rock formations of the eastern reaches of the North Cascades. Here, above six thousand feet, the blue-purple flowers are accompanied by Lomatium, Paintbrush, Valerian, and Arnica. A bit later one might come across its blooming vines in the Rocky Mountains.

Unable to grow upright of its own accord, Purple Virgin's Bower is dependent on woody plants, rocks, and slopes for support. Its vines can then expand into a network that anchors and stabilizes the crumbly cliffs and steep talus that enable its growth. The plant and these unstable formations become a system of reciprocal support.

Columbiana's strong wiry stems also contribute to small mammals and birds, for they are ideal nesting material. The flowers endure for some time, accommodating many pollen and nectar lovers.[1]

---

1   Tilford

Gesture

The flowers of Purple Virgin's Bower have four lavender sepals rather than petals. Long and tapered, they twist open from tight compact balls, their pointed tips flaring outward. Floral stems reach upward, but as flowers open they bend demurely towards the ground. Sepals are accented with darker veins or maroon that bleeds downward from the base. With their crinkled, crepe paper-like texture they resemble miniature Chinese lanterns. Reproductive parts are greenish white, and multiple stamens form a tight cluster around the pistil. By late summer the pistil's many style branches have transformed into long, silvery seed plumes, poised for long-distance wind travel.

Compound leaves are opposite, growing in groups of three. Their leaflets' shape echoes the pointed-tipped sepals. Branches also form in opposite pairs, balanced like the fourfold sepals. This expression of balance and structure is a counterpoint to the every-which-way twisting growth pattern of the vines.

Vines are strong, wiry to woody. Rather than following their own inherent direction and structure, vines conform to the rocks, slopes, outcroppings, and branches around them. When a leaf stalk contacts an object, its growth pattern alters, causing it to encircle the object. Virgin's Bower's vines sprawl along contours seeking vertical slopes and walls; they hang in mats from cliffs. High on rocky outcroppings and north faces these plants are exposed to intense weather, during which sepal tips clamp together in one protective precision point. Trailing vines can cover a large area, where the flowers, with their shy, demure ways, can be hidden. Once one flower comes into focus, many others emerge like soft jewels.

The vines and leaves of *columbiana* and other species of *Clematis* can be made into a poultice to relieve joint pain.[2] Though internal use can cause tissue irritation and internal bleeding, experienced herbal practitioners

---

2   Moore

formulate leafing vines into tinctures for migraine headaches. The plant seems to bring relief through instigating a combination of constriction and dilation actions in the brain's capillaries.[3]

Medicine Story

The Virgin's Bower shows a way of moving in and relating to the world. Each being's growth and expression is a blend of its inner imperatives and outer circumstances. Some life forms, such as Oak trees and Blackberry thickets, have strong manifestation impulses, imposing their structures on the environment. They are more shapers than responders. At the other end of the spectrum is the Clematis and other vines. Purple Virgin's Bower conforms to the contours of slopes, rocks, boughs, and thickets; its growth follows their shapes. Unable to maintain an upright stance on its own, its nature is to grow towards and secure itself to existent vertical forms. Dependent on others for its structure, it follows its surroundings. Yet as it finds shape through its environment, it becomes itself a viable structural force, securing land against erosion. The Oak and the Clematis represent two complementary ways. In a healthy community these two forces interplay in a balanced way. When the Oak force dominates, other life forms are excluded; when the vine forces take over, other life is depleted.

Most noticeable in Purple Virgin's Bower are its languid ways—its draping vines, its softly drooping flowers. Yet hidden within its loose, yielding nature is a polarity. Though the vines concede their shape to surrounding formations, they maintain the stability of these formations. Vines sprawl every which way, yet fourfold flowers and opposite branches and leaves express balance and order. Sepals open with their tips flared, but clamp together with approaching storms. The flowers, in spite of their softness, endure, even through harsh conditions. Leaf stems clasp branches

---

3   Tilford

and the flower buds are tight balls, while feathery seeds are easily loosened by the breeze, surrendering to the sky, scattering in many directions. The herbal action of the vines and leaves, the triggering of constriction and dilation, reflects the polarity of clasping and releasing observable in the plant. The rhythmic synergy of these poles sustains both the Clematis and the health of its environment. Virgin's Bower's vines and flowers serve other creatures through their flexible, yielding strength.

The patterns of growth of the Purple Virgin's Bower inspire a meditation on the life rhythms of holding and letting go, initiating and responding, asserting and adapting, solidity and fluidity. Is our natural inclination more one way or the other? With awareness we can use our inherent approach in a balanced, cooperative way. Either extreme becomes destructive, while a rhythmic dance makes for a vital organism or community.

In a culture that promotes strong self assertion and mastery, we may not take the time to attend to what is present around and within us, to feel its currents, to listen deeply. The Purple Virgin's Bower shows the way of following, yielding, and improvising. Instead of actively asserting change and solutions, we, like the vines, may need to grope our way along the surrounding contours, to be shaped by them rather than be the shapers. Instead of altering, it might be most of service to align with the wisdom of what is present. Whether we are confronting a symptom or a perceived problem, we can ask: "What is right about precisely this?" This way recognizes the inherent intelligence of nature, of life, and surrenders to this greater force in order to learn and be transformed.

We have something to learn from our inability to control, from embracing powerlessness and dependency. The growth patterns of Virgin's Bower point to the value of those who require assistance. One's personal crisis can galvanize community, eliciting caring responses. One's courage to request support can coalesce resources and structures that benefit us all.

Columbia Lewisia

Lewisia columbiana

PORTULACACEAE

...the magician reappears unannounced from a fold in the curtain you never dreamed was an opening.

— *Annie Dillard*

> **Botanical Description**
>
> A succulent perennial with numerous leaves in a basal rosette. The inflorescence is a many-flowered panicle. Flowers are small—½ inch in diameter—and have 2 sepals, 7-9 petals, and 5-6 stamens. Petals are white with pink to magenta-rose venation.
>
> **Range**
>
> Grows on both sides of the Cascades from southern British Columbia south through the Olympic Mountains and coast ranges to northwestern California, and into the Sierra Nevada. Also found in central Washington, northeastern Oregon's Wallowa Mts. and the adjacent Seven Devils area of Idaho. In the Columbia River Gorge, variety *columbiana* appears between the elevations of 100'-4400' in rocky areas of the west and central Gorge.

Habitat and Ecology

Both east and west of the Cascade Crest, airy bouquets of Columbia Lewisia grace granite ledges, serpentine areas, basalt cliffs, and rock tumbled slopes. These flowers belong to breezes and sunlight. From their well-drained perches they are a counterpoint to the dense slabs from which they spring. Flowers appear during the beginning and middle of the peak bloom of their respective habitats. Often they are accompanied by Davidson's Penstemon, Stone Crop, and various Saxifrages.

Gesture

From a distance, sprays of Columbia Lewisia flowers are dainty pink wind dancers. The plant and its environment are an interplay of solidity and airiness. A fleshy tap root anchors the plant and assures sustenance in its dry and precarious abode. First appear narrow, succulent leaves, closely packed in a basal rosette. Several rosettes clump together forming dense mats of fleshy foliage. Not only are stems a long, slender departure from this mass of horizontal growth; they space themselves around the outside of each rosette, rather than growing from the middle. Thereby, they define an opening, an empty space in the center of the plant. They surround the leaf rosettes just as the petals and stamens surround the pistil. Stems reach high—up to almost a foot—branching at the ends into a panicle of many flowers. The panicle creates a spacious arrangement of flowers, again inviting emptiness, courting the wind.

The delightful pink appearance of the flowers from a distance is created by rose to magenta veins in white to pale pink petals. The illusion of "solid" color dissolves into peppermint twist; from close up this dainty flower appears like a court jester decked in magenta stripes. Petals glisten, styles and anthers are pink, the ovary light orange, the pollen yellow green.

Just as petals appear to be a solid color from a distance, viewed from across the landscape, the stems present a solid mat of orange-pink. As the leaves mature, their green dissolves into this same persimmon color, contributing to the display. Unexpectedly, it is the plants' vegetative organs that, from afar, grace the rocks with "floral" color.

Medicine Story

The Columbia Lewisia is the magician, a master of illusions, defying expectations. The flower is a trickster, bringing levity, reminding us to think again about what appears as rock solid reality. Its pink is an illusion created by white and magenta stripes. Similarly, any color, say red, under a microscope becomes red and white spots; magnify one of the red spots and it becomes smaller red and white spots. This process can be continued with ever smaller red spots until notions of solidity melt.

Quantum physics has shown that "solid" matter is made up of tiny, ambiguous wave / particles with enormous spaces in between. These wave / particles change with observation and could be better described as fields of possibility. Similarly, through awareness practice we can experience the insubstantial nature of our thoughts and emotions, of the very beliefs that we rely on as stable foundations for our lives. Consciously and unconsciously we interpret another's behavior or presuppose the future. Body sensations that we identify as emotions arise with these interpretations. We react to what seems to be an external truth when in actuality our inner projector is spinning dramas and illusions.

The Columbia Lewisia has a kinship with rock and with air, density and emptiness. Air moves through living beings, connecting the inner and outer, past and present, all that is. In Eastern practices breath is used to master states of consciousness; dense states can be transformed into lighter ones, even into the highest spiritual experiences. Perhaps our human purpose is to be alchemists. We grapple with what seems solid, heavy, and fixed, and eventually learn to transmute our unsolvable problems into spiritual wisdom. The delicate *Lewisia columbiana*, found amidst boulders and wind, juxtaposes the two poles of density and levity. Its alchemical prowess is expressed in its ability to flourish even in serpentine rock, which contains heavy metals toxic to most life forms.

This plant reveals that obstacles may not be brick walls after all. As the *Course in Miracles* suggests: "If you are tempted to be disturbed by something, try finding another way to perceive it." Meanings, explanations, and names can lock in one perspective. Names create the illusion that something exists as a separate, predictable entity— pink, sadness, pneumonia, a flower. Thich Naht Han points out that a flower is made up of elements that are not-flower, such as sun and clouds.[1] Minerals and insects and many other not-flower aspects are part of the flower. Labels seem to create fixed constructs, and awareness helps us avoid the trap of this illusion. The assertion that we have a certain disease or problem can give it weight and authority. We may then lose touch with our own influence and empowerment and depend solely on outside experts. Shifting

---

1   Thich Nhat Hanh

our attention to the empty spaces, to the not-something that makes up something, we can cultivate lightness. As we recognize the insubstantial that creates what seems substantial and the connection of what seems separate, heaviness, seriousness, and self absorption begin to evaporate.

Columbia Lewisia points to the empty spaces. It transmutes heavy metals to delicate flowers and in so doing bestows its gifts of color, sparkle, and levity.

Jessica's Stickseed

Hackelia micrantha

BORAGINACEAE

I suspect that the child plucks its first flower with an insight into its beauty and significance which the subsequent botanist never retains.

—*Henry David Thoreau*

Botanical Description

Perennial with 5-petaled blue flowers. Each rounded petal has a white or yellow nob-like appendage (fornice) in the center. Flowers grow in open terminal clusters. Plants reach up to 3 feet tall. Stems are bunched, prickly in the margins, and have long narrow leaves.

Range

Extends south to California, north to British Columbia, and eastward to the Rocky Mountains. These plants can be found throughout the entire Cascade Mountain region as well as in the Sierras.

Habitat and Ecology

Jessica's Stickseed makes its home in open habitats, adapting to both dry and moist conditions. In the open, plant clumps spread in profusion, casting a blue-green mist across talus slopes, spongy meadows, or along stream banks. The blue flowers first appear in early summer in the foothills of Eastern Washington just as the resident spring ephemerals are fading. They contribute to the rich floral pageant that comes to life just after snow melt. As summer waxes, Jessica's Stickseed progresses upward with this wave of color, eventually reaching the high meadows of the Cascade Mountains.

Gesture

As with many plants in the Borage family, the flower of Jessica's Stickseed appears disproportionately small in relationship to the stems and leaves. Plants are tall, growing up to forty inches, with long narrow leaves, while the flowers are only about ¼ inch in diameter, and they uncoil into small terminal clusters. Supported by a deep taproot, the long stems grow in bunches, and the plant is adept at spreading through a large area. Compared to the plant's exuberant growth patterns, the blossom seems contracted. Compared to the foliage, it seems understated, restrained.

Yet the Jessica's Stickseed blossom captures attention with its charm. Buds blushed with pink and lavender open to clear blue flowers that sparkle in morning or late afternoon sunlight. The soft blue color draws us into the five rounded petals. Each petal has a white or yellow knob near its base, so that a raised and contrasting ring forms in the center of the flower. These knobs, known as fornices, guide the bee's path to pollen-laden anthers. Tidy and symmetrical, the flowers resemble small cameos, each one a hallmark of nature's loveliness and perfection. Again, the flowers contrast sharply with the rather ungainly plants with their rough, hairy stems and leaves. From a distance the ample greenery modifies the flower's pure blue, creating a blue-green effect over the landscape.

The growth cycle of plants is a rhythmic dance of expansion and contraction. Jessica's Stickseed plays with these polarities in its own unique way, expressing variations within the plant archetype; it therefore draws our attention to these rhythms. Within the archetype of the plant, the process of flowering is one of opening and expanding. During flowering the plant achieves its greatest transformation, and the forces of ascension prevail. Jessica's Stickseed is expansive in its vigorous stem and leaf growth, while the flowering process is more subdued; tight coils open into compact flowers. In its time of transformation the plant's ascending forces appear suppressed. Yet the flower's color, the true blue that is rare among flowers, mirrors the clarity of the sunlit sky. Though the flowers are contracted in size, they are emissaries of the open sky. Also significant is the symmetry and five-fold structure of the flowers. These diminutive blue stars are emblems of the infinite regenerative power of nature and the harmony and divine proportion implicit in all natural processes. (See Chapter 3.) The flower is a distillation of the underlying harmony of the plant as a whole. Less apparent in the stems and leaves, this order is brought to light in the flower.

Continuing the paradoxes of this plant, the very compact fruits of Jessica's Stickseed express expansion in their dispersal strategy. The fruits are four small nutlets ringed with curved barbs. These nutlets capture the attention—as well as the clothing—of humans so persistently as to inspire the plant's common name. This adaptation is a rather aggressive one, allowing the plant to spread wherever its seed transporters may wander.

Medicine Story

Expansion and contraction, dispersion and concentration, opening and closing are life rhythms. These movements underly the mysterious process of creation. The dance of Jessica's Stickseed speaks of creative manifestation. In its growth habits, the plant is bold, willing to take up space. In the flower, restraint, precision, and distillation come into play. When these two poles are in gentle and harmonious relationship, creative expression gracefully unfolds. Through its flower's deference to restraint, Jessica's Stickseed gives the appearance of an imbalance of these polarities. Like the wounded healer, the plant's medicine is revealed in its "wound". Jessica's Stickseed encourages us in exuberant creative actualization.

The soft colors and simplicity of the Jessica's Stickseed flower express a childlike quality; this flower recalls the innocence and purity of childhood. Reasonably healthy children naturally undertake creative endeavors with abandon. Unmindful of censure, they are willing to make a mess, to become totally absorbed in their process. Many of us remember such a time of delight and freedom.

Jessica's Stickseed invites us to consider the ways we hold self expression in check. This plant offers medicine for reclaiming spontaneity, for allowing the natural unfolding of our unique creative potential. In Jessica's Stickseed we encounter the pure blue of the open sky, and its little flower reconnects us to the unrestrained delight of a child's creative play.

Wyethia

Wyethia amplexicaulis

ASTERACEAE

The world is always turning towards the morning.

__Gordan Bok

Botanical Description

One or more large, composite flower heads bloom at the ends of long erect stems. Ray flowers are deep yellow, disk flowers lighter yellow. Basal leaves are long; stem leaves smaller, alternate, and sessile. Foliage is entirely smooth and shiny with resinous texture. This is the only Wyethia species in Washington with smooth (hairless) leaves and stems. Grows up to 30 inches tall.

Range

Grows in eastern Washington, eastern Oregon, and in the more northern states of the intermountain west, including Nevada, Idaho, Montana, Wyoming, Utah, and Colorado. In the intermountain west it grows from 4500 to 11,000 feet elevations.

Habitat and Ecology

Wyethia, sometimes referred to as Mule's Ears, grows on the east side of the Cascades from the foothills to the middle elevations of the mountains. It flourishes in many plant communities, but is most prolific amongst sagebrush in the upper shrub steppe. In late spring and early summer it brings sunny splashes of color to open woodlands, forest edges, and moist draws. On open slopes and springtime wet meadows it can amass up to 2700 plants per acre![1] These plants often share habitat with the better known Balsamroot, whose slightly larger floral pinwheels have just begun to fade when the first Wyethia blossoms are appearing. It is as though the golden torch has been passed, ensuring this land will be ablaze a bit longer with sun-like beauty. Since its blossoms join the peak season flowers late in their display, wherever Wyethia grows it extends the climax of color.

---

1 http://www.cwnp.org/photopgs/wdocymamplexicaulis.html

Gesture

Wyethia's inflorescence is characteristic of many plants within the Asteraceae family. At the same time, it has qualities that distinguish it from these relatives.

In the Asteraceae family what appears to be a single flower is actually a community of many small flowers, sometimes referred to as florets, which are of two types. The petal-like forms that extend outward are known as ray florets, while the tubular shapes that cluster together are disk florets. Wyethia, as a representative of the Sunflower tribe of the Asteraceae family, integrates both types of flowers into its large, round inflorescence. The ray florets radiate around the edge while the disc florets form the center. Each blossom is like a whole meadow of flowers. Each is a clear example of a holon, something whole unto itself, which is made up of smaller wholes, and which is also part of a greater whole. The composite flower is part of the greater whole of the plant, which in turn comprises a plant community. Wyethia and other Sunflowers are emblems of the holonic nature of the universe.

As we gaze into a Sunflower blossom, it seems to radiate out while also drawing us into its center like a mandala. The blossoms of Wyethia, like many other Sunflowers, are bright yellow. In shape, color, and brightness, they imitate the sun, mirroring back its glory. The sun, like these blossoms, possesses both a radiant outward force as well as a magnetic pull towards its center. Radiance, magnetism, integration, oneness, and wholeness are part of the solar archetype. The sun's magnetism draws planets into orbit and lifts plants towards the sky.

Sunflower blossoms typically follow the path of the sun across the sky from east to west each day; they respond to the sun's magnetism, making it tangible. However, Wyethia's inflorescence, in a gesture that is rare among Sunflowers, fixes its focus in one direction—the place from where the first rays of the morning sun emanate to the plant. Throughout the day and night its blossom maintains this eastward orientation. In a setting with shaded areas, where the eastern horizon is blocked by rocks, trees, or mountains, this effect is diffused. In an open area, where Wyethia has spread into dense, expansive communities, eastward facing flowers create a dramatic effect. In the slanted light of morning and evening the effect is heightened. At these transition times of day a meadow of blossoming Wyethia glows with brilliant light. This breathtaking sight is a result of the plant's characteristic response to the magnetism of the *rising* sun.

The leaves and stems of Wyethia are varnished and resinous. Like Balsamroot, this plant has balsamic properties, which can be detected in its aromatic resin. The warm, soothing, and stimulating effect of the resin is another way the plant radiates sun-like qualities. Elk, mule deer, and other grazing animals cannot resist the fragrant flowers. Native people had methods of preparing the root both to access its medicine and to enhance its sweetness as a food source.[2]

Early in Wyethia's growing season the long taproots suck large amounts of water from their meadow environment. This provides them with ample moisture and also excludes other species, assuring Wyethia's dominance. Again, this calls to mind the sun, which as it waxes draws the stores of water from the land. The sun, the bringer of green life, at the height of its dominance becomes life's fiery transformer. Early in the season Wyethia's young succulent leaves appeal to mule deer and bear, but by midsummer in the lower portions of its range, the foliage turns course and leathery. At this time the extreme dryness of its habitat is echoed in its brown flower heads. The plant's vital force is now concentrated in seeds ripening for grouse and ground squirrels.

---

2   *Ibid.*

Medicine Story

At times we lose sight of our life's direction and purpose. We may be pulled off course by conditioned reactions to pain or by cultural norms. Consider a time of darkness in your life when you lost a sense of meaning and direction. This may have been a time of intense suffering when you believed light would never again touch your life.

The Wyethia, its blossoms ever turned eastward, is an affirmation of life's renewal through light's eternal return. It embodies faith in the orderly turning of planets. The eastward direction, the direction of the rising sun, is associated with new beginnings, resurrection, inspiration, and spiritual awakening. Another word for east is orient, and the east is connected to the power of orientation, or the ability to align with a guiding principle. With the rising sun we are reawakened, renewed, and inspired, and we have the opportunity to reorient ourselves to our life course. For each of us there is a special alignment—on the Earth, within a community—which allows our own creative spark to shine. From this place we draw to us what honors our integrity and serves our life purpose. Our true orientation also helps maintain the integrity of others and the greater community. Our uniqueness is reflected in the specialness of each other being, and the more each one shines the more radiant we become, enhancing the brilliance of the whole.

When we cycle through times of darkness we often experience ourselves as isolated and alone. Wyethia is a beacon that assists our return to the unique truth of who we are. From here we recognize our integral connection to the whole of creation.

Tweedy's Lewisia

Lewisia tweedyi

PORTULACACEAE

What is the purpose of life but to make light visible? And to be dazzled by its every revelation.

> **Botanical Description**
>
> Plants have rosette of succulent leaves. Leaves are shiny, oval, entire, and spoon-shaped; they are 3-6 inches long, with petioles of the same length. Flowers are large and showy; 1-3 grow from 4-8 inch stems. Petals 10-12, apricot-pink to creamy white, stamens many.
>
> **Range**
>
> Grows only in the Wenatchee Mountains on the east side of the Cascades in north central Washington and in an adjacent area of British Columbia.

Habitat and Ecology

Endemic to Eastern Washington's Wenatchee Mountains and a small adjacent area in British Columbia, Tweedy's Lewisia is a rare plant! Its niche is this one area of scant precipitation and well-drained soils where winter brings a protective cover of snow. Here Tweedy's Lewisia grows at low to mid elevations on rocky slopes and cliffs, from rock crevices, along road banks, and in dry open forests. It favors granitic talus as well as the duff of Ponderosa Pine. Often it grows in semi shade, places infiltrated with strong sunlight. Its flowers may appear in an open meadow accompanied by blooms of Paintbrush, Balsamroot, and Lupine. Mid to late May can be a good time to look for it at two to three thousand foot elevations.

Gesture

The petals of Tweedy's Lewisia are luminescent. Washed with shimmers of color, edges of rose melt into peach, into lemon, into glistening green centers. A spray of bright orange anthers completes the presentation of this light-filled flower. Stem tips droop slightly under the weight of open blossoms. These flowers' ample size combined with their ethereal glow will stop in one's tracks anyone fortunate to come upon them.

Leaves are also large and grow from leaf stems equal to their length. They are oval and curved inwards at the edges, perhaps to funnel drops from a spring rain shower into the heart of the plant. Fleshy, luxuriant green, and packed into a basal rosette, the leaves add to the plant's showiness. The leaves' succulence and the plant's thick taproot, which reaches deep into the ground for moisture, helps Tweedy's Lewisia adapt to its dry environment.

Medicine Story

According to spiritual teacher Eckhart Tolle, flowers can be understood "as the enlightenment of plants." Tolle explains that flowers are almost beyond material substance, "more fleeting, more ethereal, and more delicate than the plants out of which they emerged... like messengers from another realm, like a bridge between the world of physical forms and the formless."[1] They mirror our light-filled essence, which transcends our physical bodies and personalities. Flowers open our hearts with joy as we recognize this divine essence that we share with all life. Perhaps ever so slightly or perhaps in a very palpable way, an encounter with a flower furthers our human evolution. A blossom allows this glimpse behind the illusion of solidity and separation just as do crystals and gems, birds, rainbows, and new born life.[2]

What amazing materials comprise a flower! No wonder floral colors and fragrances are compared to divine emanations. The petals of Tweedy's Lewisia seem to be woven of light itself. The luminescent flowers of this plant highlight the special gift of all flowers. Tweedy's Lewisia plants express the soul's alchemical journey through dense substance to pure light. Their fleshy taproots anchor them in rocks and hardpan soils; shiny green leaves are thick, retaining water; and the large flowers open into shimmering light and color. This expression is similar to Water Lilies,

---

1   Tolle, p. 3

2   *Ibid.*

whose growth from mud through layers of water of increasing clarity culminates in the flower's radiant opening in air and sunlight. In Eastern traditions the Water Lily has long represented the soul's journey to enlightenment.

As with the Water Lily, both dense matter and translucent flowers are part of the growth process of the Tweedy's Lewisia. Similarly, the human journey of awakening to the light of unity consciousness requires embodiment. The material world and dense emotions and thought forms are the ground and substance from which we forge spiritual growth. We are of greatest service to each other not by dwelling continuously in the light, but by again and again grappling with heavy states of consciousness, transforming them into unconditional self love and compassion. Each time we emerge from the darkness we are able to offer forgiveness, hope, and possibility. The Tweedy's Lewisia is a celebration of the connected circle of involution and evolution, of the precious gift of embodied life on Earth through which light is made visible.

My first encounter with Tweedy's Lewisia blossoms occurred while trudging off trail up a steep forested slope under heavy backpack. Unexpectedly, like a vision, they appeared. At once the weight of my load and the uphill struggle transformed into wondrous discovery. I could only drop to the ground, lost in the translucent rose, peach, and green for which there are no adequate names. The land around me, stark, harsh with rocks, took on a new meaning. It held the precise factors that converged in the possibility of such a plant.

Beargrass
Xerophyllum tenax
LILIACEAE

"I Imagine Us As a Holy Family Engaging In
the Great Work of Increasing the Light"
—*Cappy Thompson (title of her stained glass mural
on the front of The Evergreen State College Library)*

Botanical Description

Numerous creamy-white flowers cluster in a bulbous, elongated inflorescence (up to 20 inches) from ends of sturdy, erect 1-5 foot stems. Six tepals with longer stamens. Leaves are evergreen, narrow, and tough, edged with minute barbs; stem leaves smaller, basal leaves long, 1-2 feet, growing in grass-like clumps.

Range

Beargrass grows in the Cascades and coastal mountains from British Columbia into Northern California, where it is found in the northern part of the Sierra Nevada Mountains. Its range extends east to the Rockies as far south as Wyoming.

Habitat and Ecology

Shiny, dark green, grass-like clumps are familiar to high country seekers as they ascend through mid elevation conifer forests. These clumps are the long, wiry basal leaves of Beargrass, which grows under the denser canopies of west side forests as well as in the drier, more open woodlands east of the Cascade range. In drier forests, this plant can completely dominate the understory. As trails wind to ridge tops and forest openings, one begins to notice flower stalks, perhaps blooming, arising from these clumps. Subalpine meadows might greet the hiker with striking displays of these stately flowering plants. The southern Cascades, where volcanic activity has deposited layers of pumice, often host vast communities of Beargrass. Other plants can't easily take hold in these rapidly draining pumice soils, but Beargrass, with its dry, tough leaves and moisture garnering clumps, can thrive here.[1] Statuesque "forests" of this plant may also be encountered in burn sites. In the west Columbia Gorge area, Silverstar Mountain—where a great fire swept the land in the early twentieth

---

1 Mathews

century—hosts impressive stands of Beargrass on its high, open ridges and flanks. Beargrass seeds and plants survive and benefit from exposure to the conditions of low level fires.[2] It is among the second wave of plants—after Fireweed—to enter a burned area. It is not able to grow immediately after a high intensity fire in the way of Fireweed, but it plays a role in the succession of restorative plants.

Beargrass grows in very different conditions and soil consistencies, and its absence or prevalence in an area can be puzzling to predict or understand. With its ability to thrive in dry soils, Beargrass communities are widely found east of the Cascades and into the Rocky Mountains. Yet it also does well in wet west side soils, such as sphagnum or clay, a sharp contrast to its dry well-drained sites. Beargrass thrives in open sun-filled places such as west and south facing slopes and ridges, while it also adapts well to shade, where it often out competes other plants. It is a middle to high elevation plant, growing into alpine habitats. Yet there are two areas where low elevation Beargrass sites occur: the Olympic Peninsula and the Oregon Coast.

The blossoming time of Beargrass is in May at low elevations. On the flanks of 4,000 foot Silverstar, flowers peak around summer solstice if snow pack and spring temperatures have been moderate. Beargrass' bloom time extends into July and August in its higher and more northern abodes. Plants develop for several years before they blossom, longer in shaded areas, so Beargrass communities follow cycles of flowering. In certain years a habitat, such as the high ridges of the Goat Rocks, treats us to breathtaking displays of hundreds of blooming plants, while the next year flowers may scarcely be found.

Rodents, pikas, and larger mammals feed on various parts of Beargrass plants. In springtime, leaf clumps have succulent bases, and one theory suggests that bears feeding on these inspired the plant's common name.

---

2   Shebitz

Early in the season when new, still somewhat tender leaves are growing from the clump and other food sources are scarce, deer may graze on these. Pikas also partake, perhaps harvesting some to dry for winter storage.[3] Only Rocky Mountain goats perennially feast on the harsh, sharp mature leaves. Since leaves remain all year, they are crucial winter forage for these goats. Elk and other grazing animals favor the flower stems, in bud or full flower. Certain species of mice eat stem bases, thus killing the plant.[4]

Its long, pliable and highly durable basal leaves have made Beargrass invaluable to Northwest Native people. For many generations they have woven completely water tight baskets from these leaves. Thankfully, this fine art continues today. Northwest Natives also wove Beargrass leaves into capes, hats, mats and other clothing and artifacts.[5] The leaves of Beargrass were used in the burial of influential people[6], an expression of reverence for this plant and its bestowal of valuable gifts.

Gesture

The long, tough basal leaves are an important part of the gesture of Beargrass. Their dense evergreen clumps, ever conspicuous, connect this plant to people, whose survival they have enhanced. Beargrass' botanical and common names arise from the qualities of its leaves and their human application. *Xerophyllum* translates as "dry leaf", and *tenax* means "tenacious", or "holding fast". Other common names for this plant are Squawgrass and Basketgrass.

Basal leaves are narrow and pliable as well as durable, tough, and long, growing up to thirty inches in length. These attributes, valuable

---

3   Mathews
4   Craighead et al.
5   Mathews
6   Shebitz

for crafting artifacts, also imbue the plant with endurance and survival capabilities. Leaf edges have minute barbs and can wound fingers or soft animal mouths. Leaves are quite slippery under foot; they are glossy, reflecting light even under cloud cover or forest canopy. In sunlight Beargrass clumps gleam with silvery radiance. The leaves' durability, sharpness, slipperiness, and sheen indicate the presence of silica. This crystalline substance is also apparent in grass and conifer needles.

Beargrass grows from thick woody rhizomes, which enable its efficient spreading. The basal leaf clumps develop slowly and it takes years before the plant sends up flowering stalks. The plant must accumulate an ample amount of nutrients from photosynthesis before it flowers. In shady locations even more years must elapse before stalks appear, and some of these places see few blooms. The sturdy, erect flower stalks can grow from three to almost six feet tall. Small wiry leaves spiral up each stalk. These stem leaves grow progressively shorter as they approach the flower cluster at the top. Once the plant flowers and seeds have developed, the whole plant dies, and it funnels its remaining nutrient stores to a new plant through its rhizomes.[7]

At the upper part of each stalk many small, creamy flowers combine to form distinctive white plumes, wider in the middle than at the ends. These floral clusters open from the bottom up, the yet-to-open top flowers forming an upward pointing nipple. As flowers develop, this inflorescence slowly elongates, and it can reach twenty inches in size. This bulbous plume forms from ½ inch flowers that each grow from a short stem. The flowers' six tepals (identical petals and sepals) open wide into radiant little stars. Stamens protrude beyond the tepals, giving a furry texture to the overall flower plume. Flowers are more musky than fragrant, another characteristic offered to explain the name "Beargrass".

---

7   Mathews

Medicine Story

With their glistening leaf clumps and white plumes atop regal stalks, Beargrass plants are sentinels of light. They take in and transmit light, magnifying its presence. Rather than flowering each year, Beargrass devotes its entire life span to a final season of glory. Several years of stored sunlight at last propels stalks skyward and opens clusters of starry blossoms. Beargrass' torch-like presence transforms the land. Then, as their radiance fades, plants bestow what remains on the next generation.

Beargrass expresses ascension in its gesture. Life responds to light by quickening and reaching upward toward its source. Walking among flowering Beargrass, we are energized and uplifted by its stately vertical stalks and airy light-filled plumes. Its sixfold flowers and silica rich leaves and stalks reveal the plant's kinship with crystals and water. Crystalline structures, whether of water or minerals, are hexagonal, and they magnify and reflect the brilliance of light. This relationship to light is connected to the ability of these structures to receive, store, and transmit energetic patterns—that is, communications—to living tissue.[8] Beargrass medicine helps us receive light and be nourished by it. Physical light raises our energy level and thereby lifts us to a higher state of consciousness. Physical light is like a message to our whole being reminding us of the magic and beauty of life. This inspiration dissolves dark, heavy states, and opens us to receive more spiritual light. We then remember the brilliance of who we really are.

Beargrass communities thrive at high elevations. The mountain goats that inhabit the rarefied, craggy reaches of the Rockies depend on the evergreen leaves of Beargrass for a continual source of nourishment. The plant's relationship to the high mountains and its creatures again reveals its sublime qualities, its connection with the spiritual realm and higher states of consciousness. Beargrass' evergreen leaves relate it to what is eternal, the truth and wisdom that transcend space and time.

---

8   Greene; Ryrie

Though Beargrass expresses light and ascension, its gesture is also feminine. It carries the medicine of a wise, elder grandmother—perhaps a grandmother who is tough and perseverant with sharp edges. She knows the way of survival and nourishes the life around her. She teaches us about resilience, endurance, and tenaciousness. This grandmother is a keeper of the flame of life. The flower clusters of Beargrass are milk-white and breast-shaped. Its long leaves clothe and sustain the people. Through its network of rhizomes it nourishes new plants, and it restores the land after fire and destruction. Rather than a sublime fragrance, the flowers have a musky scent, an expression of the plant's earthy quality. With its spreading rhizomes, Beargrass is a plant of community, and this is reflected in its clusters of flowers and clumps of basal leaves. Its floral plumes are more lunar than solar in their color, and a Beargrass meadow in moonlight is a sight to behold.

Beargrass, like other plants, is a bringer of gifts of light. Through photosynthesis its leaves transform sunlight into physical nutrients accessible to living creatures. The leaves provide raw material for physical survival needs. Through its flower, the physical nutrients are transmuted back to the ethereal; the radiance of the flower is light returned to light. We are nourished by light both through the denser form of carbohydrates and through the fast pulsating energy of direct sunlight. Like direct sunlight, flowers—along with water, crystals, and nature's clear, radiant colors—are high frequency sources of pure light. This light feeds both our cells and our souls, as it uplifts us and opens us to the nourishment of spiritual light. Like Beargrass you are garnering lifetime stores of spiritual light. Exposure to light and nature's beauty will nurture the great flowering of your life. The more you absorb the beauty of the flowers and all that you love, the more light you have to pass on from your life's luminous torch.

Fireweed

Chamerion Angustifolium
ONAGRACEAE

It is only by acknowledging impermanence that there is a chance to die and the space to be reborn and the possibility of appreciating life as a creative process.

_Chogyam Trungpa

> **Botanical Description**
>
> Native perennial with narrow lance-shaped leaves that alternate up 3-9 foot stems. 15 or more rose-pink flowers grow in a spike at the top of the stem. Sepals 4, Petals 4, Stamens 8; inner 4 stamens are shorter than outer 4. Ovary inferior and 4-chambered. Fruit is a capsule with 300-500 seeds. Seeds have a tuft of long hairs on one end. Fine roots with rhizomes extend as far down as 18 inches.
>
> **Range**
>
> Circumboreal; in all of the Canadian provinces and throughout the United States except in the southeastern states and Texas.

Habitat and Ecology

Fireweed is almost universally known in the temperate regions of the Northern Hemisphere, for it grows from the lowlands to the high mountain meadows, in wilderness ecosystems as well as in cultivated land and vacant lots. It is a pioneer plant, one of the first to establish itself in habitats that have undergone some form of destruction. In the wild, Fireweed springs to life after fires and along the slopes and basins of avalanches. It also follows the scars of human industry, spreading through the stumps and slash of logging sites, along roads and railways, and amidst the blackened debris of urban fires.

Fire sets the stage for the densest and most far-reaching stands of Fireweed. History tells of the waves of magenta that rippled through the charred remains of London following the air raids of World War II. Areas such as the burned portions of Yellowstone National Park ignite in brilliant color during Fireweed's blooming time.

Fireweed blossoms are long-lasting, and flowers begin with the onset of summer in the low country and continue into late summer in the mountains. When mountain slopes are colored with autumn, clumps of silken seeds gleam amongst stands of Fireweed. As the breeze tugs at them, one by one they swirl upwards, and the limitless sky is filled with white parachutes. This spectacle explains the wide and varied distribution of Fireweed plants.

While Fireweed is part of the first stage of recovery in a logged or burned forest, in a few years it will begin to be replaced by the next wave of plants, including Beargrass and Huckleberry. Where there has been prolonged fire suppression, excess debris fuels intensely hot fires, and Fireweed is one of the few plants that can establish itself in their aftermath. Fireweed may grow in such areas for many years before other species find them hospitable.

Gesture

Fireweed is striking for its vivid flowers, whose blooms swirl upwards on their narrow raceme like a magenta torch. Flowers have parts in fours: embraced by four cranberry-colored sepals, petals are arranged in a Celtic cross; stamens are eight and the white pistil has a four-pronged stigma. Anthers bear blue-green pollen, a lovely contrast to the rosy petals.

Fireweed has long, slender seedpods that ripen below the sepals until they burst, releasing copious downy seeds. The seeds' ability to ride the winds for long distances and to remain viable for decades combines with the plant's ability to spread quickly through rhizomes to make Fireweed extremely opportunistic. Widely dispersed seeds can lay in wait for a disturbance to occur, and, once germinated, plants rapidly sweep through an area, colonizing it to the full extent. Fireweed has a strong horizontal gesture; it extensively covers the four directions of the Earth. At the same time, the plant is a tall, usually unbranched spike with its flowering racemes reaching skyward. Narrow lance-like leaves spiral alternately up the vertical stems. Their shape echoes the upward spire of the overall plant. With both a strong horizontal and vertical gesture, Fireweed imitates fire itself; it rapidly overtakes large swaths of land while its flame-like leaves and spikes reach for the sky. Its soaring seeds are not unlike the ash and sparks of a forest fire.

Afternoons find meadows of Fireweed alive with bees, gathering the flowers' abundant nectar for delicious Fireweed honey. During twilight times deer and elk come to browse. Native peoples have enjoyed the tender shoots and young leaves as well as the soft pith inside the stalks.[1] Fireweed stands are magnets for life, offering vivid color, sweetness, and nourishment. While Fireweed is restorative to fire-swept habitats, its leaves and flowers have anti-inflammatory properties for humans and animals. Taken as a tea or applied topically as a wash, the leaves and flowers release compounds that cool and soothe inflammation in internal organs, body orifices, and injuries or skin conditions.[2]

Medicine Story

Because it is able to establish a vigorous foothold in a damaged ecosystem, Fireweed is a plant of renewal; it is one of the first species to begin a process of recovery. Even where the life forces of an area have been completely destroyed, it can establish itself and thereby weave a new matrix for life. Its network of roots that restabilizes the soil is one way Fireweed supports life's fragile reentry. The plant's contribution to life's renewal operates on both concrete and subtle levels. The fourfold geometry of its flower symbolizes solidity and support. Four is the number of matter, the mass and structure of the sacred Mother. It represents the Earth with her strength and capacity to hold and nourish life. Physically and energetically Fireweed provides a new structure that can anchor and nourish a succession of life processes. Its magenta flowers uplift and reinvigorate the spirit; they attract new life with their beauty. Their color is of high frequency, at the edge of the visible spectrum. It relates to the

---

1 Pojar
2 Moore

violet flame with its power of transmutation, the ability to transform substances to a higher form.

The medicine of Fireweed can assist us when the fires of change have swept through our lives. When we are devastated by loss, there are seeds that can germinate at just these times. These seeds, like those of Fireweed, have laid in wait for years or decades for just the right conditions. In the collective human species there are seeds of wisdom that have remained viable for eons, able to persist until the one fortuitous moment of necessity that will ignite them into life. Fireweed reflects to us the miraculous process of renewal that emerges against all odds, that is part of the wisdom of both nature and the psyche. This beautiful flower helps us contact the new life-giving matrix that is forming both within and around us. Through Fireweed we can notice that there is beauty in fire, that loss is form changing into new form, that transformation is life ever transcending itself.

Fireweed's ability to inhabit and rejuvenate places that seem hopelessly devastated makes it a living emblem of the resilience of nature. It is easy to lose faith in this resilience as our technology of violence becomes ever more advanced and as we witness the sobering effects of our disharmonious ways of living. Personal and collective trauma can likewise cause disbelief in the ability to recover, disbelief in the goodness and beauty of life. Fireweed medicine soothes and transmutes what seems unbearable. It embodies the power of renewal and resilience inherent to nature and to the human psyche, and the keynote of this power is vibrant color and beauty. Fireweed affirms that, paradoxically, suffering and devastation, even the shattering of a life, can trigger the emergence of unexpected beauty.

Common Paintbrush
Castilleja miniata
SCROPHULARIACEAE

The body will again become restless
Until your soul paints all its beauty
Upon the sky.

—Hafiz

> **Botanical Description**
>
> Perennial herb 12-24 inches tall with entire, lance-shaped or linear leaves. Flower bracts are pink to scarlet-red or bright yellow-orange. Green flower beaks extend beyond bracts as they mature.
>
> **Range**
>
> Grows in the mountains of western North America, from Alaska through British Columbia and the northwestern States into New Mexico, Arizona, and southern California. It is missing only from the coast ranges of California and Oregon.

Habitat and Ecology

What plant genus could be more emblematic of the North American West than *Castilleja*, or Paintbrush? In summer Paintbrushes emblazon wild landscapes with every hue of a big sky sunrise.

The *miniata*, or Common Paintbrush, is considered the most prevalent species, growing throughout the western US and Canada in a wide variety of habitats. It is found from low to high elevations, from the coast to the mountains. It frequents wet areas such as moist meadows and stream banks, but also thrives in dry prairies. It is abundant in open, sun-filled places, but often spreads beyond forest edges into deeper shade. One may encounter Common Paintbrush along roads and trails, in tidal marshes, thickets, gravel bars, grassy slopes, and subalpine meadows. In the Columbia Gorge it grows at middle to high elevations on both the Washington and Oregon side as far east as Lyle.[1] Look for it at Silverstar, Table, and Larch Mountains and at Multanoma Falls.

All Paintbrushes are partially parasitic, drawing a portion of their water and nutrients from the roots of grasses and a wide variety of flowering plants. The Paintbrush extends specialized roots that contact then

---

1   Jolley

penetrate the roots of other plants, extracting some of their sustenance.[2] This reliance on their surrounding plant communities makes Paintbrushes very difficult to transplant, a trait that may protect them from collectors.

Gesture

While the Paintbrush genus as a whole makes a vivid impression on our memories and powers of recognition, even experts have difficulty discerning individual species. The genus traits are striking, while attributes delineating species are variable, difficult to observe, and confusing because of hybridization. Some species names—both common and botanical—refer to color. However, the same species may appear in a wide variety of colors. For example, a predominantly red species may also grow in shades of orange, purple, and pink.

As a genus, Paintbrushes appear in almost every color of the rainbow. While the signature Paintbrush drips with vibrant crimson, plants display every red-related hue as well as yellow, cream, white, and apple green.

To understand Paintbrushes, it is helpful to begin with the characteristics of their family and genus. The Figwort family—known as Scrophulariaceae—is comprised of irregular flowers, ones with bilateral rather than radial symmetry. Petals are fused into a corolla tube with two "lips", the upper with two lobes and the lower with three. These flowers are advanced in their pollination strategies; each species' size and shape are a custom fit for one or very few pollinators.[3]

Paintbrushes—at least the ones displaying colors related to red—are pollinated by hummingbirds. In the Paintbrush genus the brightly pigmented parts are actually modified leaves called bracts. Concealed among them are the flowers, which are inconspicuous in size and color. The bracts display the colors that we associate with flowers, while the flowers

---

2   Biek; Craighead et al.

3   Mathews

are undistinguished shades of green. The bracts lure the hummingbird with bright pigments, while the flowers' long, narrow corolla tubes perfectly accommodate its bill. A hand lens can help you understand the flower. Two fused petals comprise the upper lip of the corolla tube. Known as a "galea", it houses the four stamens and forms a beaked tip. The stigma sticks out beyond the galea's beak. The corolla's lower lip is short, sometimes reduced to three stubby knobs.

Paintbrush leaves grow alternately on erect stems, and they are either entire or divided into lobes, a distinction that helps with species identification. Even though their leaves produce chlorophyll, Paintbrushes' parasitic relationship with other plants supplements the nutrients they produce from photosynthesis.

The *miniata*, or Common Paintbrush, may be distinguished from other species by its size. Also known as Giant Paintbrush, it grows robustly with branching stems and reaches almost three feet. However, in higher zones it is smaller, as short as eight inches, and more often unbranched. Its leaves are important for identification, for they are typically entire, that is, not divided. They are flat and linear or lance-shaped. The ovate bracts are also undivided, but can be tipped with short, sharp teeth. The beak of the *miniata*'s galea is slightly hooked and protrudes beyond the bracts. Erect stems, smooth or with short fine hairs, sometimes sticky, arise from woody root crowns.

Careful, detailed observation is required to identify the Common Paintbrush. Yet confusion is still possible thanks to their traits' variability and hybridization. Color is no help. Even though its species name, *miniata* refers to "minium", a scarlet-colored oxide of lead,[4] *Castilleja miniata* appears in many shades of pink, bluish-red, bright yellow-orange, and pale green, as well as scarlet. Lovely rose-colored plants grow in abundance in British Columbia's Chilcotin mountains.

---

4   Pojar; Clark

The coastal Makah people employed scarlet-bracted Common Paintbrush to lure hummingbirds into traps. Hummingbirds were sought for charms in whale hunting.[5]

The flower essence of *Castilleja miniata* helps one sustain intensive creative output by grounding and revitalizing the physical body. It rouses the life force to enable one to bring inspirations into physical form.[6]

Paintbrush plants have astringent properties and have been used to stem the flow of menstrual blood, as well as to assist with other women's issues. Rheumatism has been treated with Paintbrush, and the flowering tips are considered edible in small quantities.[7]

Medicine Story

A plant's departure from usual plant forms reveals its medicine. In Paintbrush, bracts—which are leaf structures—are bright with pigment, while flowers are inconspicuous, even hidden. Flowers are unremarkably green, like foliage, while bract colors are especially vibrant. It is as though the plant brims with life force, which, with unabashed exuberance, spills onto the upper foliage before it reaches the usual main attraction—the flower. Paintbrush is also unique because even though its foliage produces chlorophyll, it is dependent on other plants for sustenance. Both unusual qualities draw attention to life force energy, its creative expression, its depletion, and its replenishment.

To our imagination the plant spikes are artists' brushes dipped in bright pigment. Through many aspects of its gesture, Paintbrush reflects vibrant artistic expression. Dripping with color, it elicits our longing to create. It shows us how to balance and sustain creative energy, drawing sustenance from the ground, from the community around us. The upright

---

5   Mathews

6   Kaminski, Katz

7   Elpel

plants seem to align fire with earth, bringing vision into form. They tap Earth's living systems and offer back flames of color.

Paintbrush spreads a feast table of colors, each with its own gifts. The fiery hues of Paintbrush generate palpable physical energy. The rose-colored plants are more sublime. They suffuse their surroundings with a delicious, heart-opening quality. Like sacred works of art, they inspire higher spiritual states. The green and cream-colored Paintbrush are subtle, accentuating the colors of other meadow flowers.

Paintbrush activates us to *create*, to imbue the world with many colors. It invites us to extend hungry roots into our surroundings, to overflow, matching its exuberance with our beautiful works.

Shrubby Cinquefoil
Potentilla fruticosa
ROSACEAE

All sanity depends on this: that it should be a delight to feel heat strike the skin, a delight to stand upright, knowing the bones are moving easily under the flesh.

__Doris Lessing

> **Botanical Description**
>
> Woody shrub with leaves divided into 3-7 pointed leaflets with silky hairs. Flowers, with 5 bright yellow petals and many stamens, grow from twig ends.
>
> **Range**
>
> Circumboreal.

Habitat and Ecology

Shrubby Cinquefoil grows in diverse habitats, and is widely dispersed in temperate and northern latitudes. It is found in the dry shrub steppe as well as in the tundra's soggy muskegs. In urban areas, it is a popular commercial landscaping plant, and humans may or may not notice its sunny flowers as they bustle into office buildings. It is most radiantly beautiful, its flowers most brilliant, in the subalpine and alpine regions where it grows along exposed ridges and talus slopes. In these knarly, harsh environments, it contracts into short bushes and misty green mats, yet it tends to be among the largest of the flowering plants here. In the alpine, Shrubby Cinquefoil blooms with Paintbrush, Prairie Lupine, Cascades Penstemon, Alpine Buckwheat, Stonecrop, and white Saxifrage. For a few precious days in midsummer these talus slopes are a Mexican fiesta.

Gesture

Typical of members of the Rosaceae family, Shrubby Cinquefoil is not shy. Its vigorous growth occupies more space than its alpine neighbors. It brings a big splash of color to the landscape, for its shrubby mounds are laden with butter yellow flowers. Each flower expresses the vibrancy of the overall plant. Five glowing petals curve delicately around a sunburst of stamens and a glossy, multifaceted stigma. A five pointed star shines from the flower's center, as yellow-green bracts peer through separate petals.

From rough and prickly woody stems grow soft fleshy leaf petioles. Compound leaves mainly have five leaflets, whose silky hairs give a misty, gray cast. Almost needle-like, leaflets resemble tiny fingers of partially closed hands. With their hairy softness and narrow, folded gesture, they guard precious water from alpine sun and wind.

Wind, scree, fine dust, heat, and icy cold sculpt the alpine and subalpine Shrubby Cinquefoil, who perseveres, growing by increments. In the high montane the larger mounds and small bushes are ancient plants, some over a hundred years old. They are worthy of the respect of elders.

Medicine Story

Perseverance and regenerative power characterize the Shrubby Cinquefoil. Even through harsh conditions, it possesses a joie de vivre; not only does it survive but it flourishes, its flowers offering back the brilliance of the sun. This shrub thrives in many habitats, whether marshy or extremely dry, in high and low elevations, and in tundra. It is as though it wants to experience as many of Earth's environments and combinations of elements as possible, absorbing experiences and contributing its own sensual delights.

Shrubby Cinquefoil demonstrates how to take root, take up space, and shine. Its yellow flowers uplift the spirits, for they generously reflect the sun's radiance. The flowers imbue their surroundings with light. Shrubby Cinquefoil teaches willingness to be vibrantly alive, reveling in Earth's sensual treasures.

The gifts of sunny color and adaptability combine with the plant's other offerings that characterize all members of the Rose Family. As an emissary of the Roses, Shrubby Cinquefoil carries the power and beauty of fivefold geometry. The center of its flower is emblazoned with a five-pointed star. This star carries the twin mysteries of the golden mean and the Earth's inexhaustible, self-sustaining regenerative power. (See Chapter 3.) Each Cinquefoil flower is a cipher pointing to the beautiful harmony and infinite life-giving capacity at the core of Earthly life. The message is, at the deepest level, all is well.

River Beauty

Chamerion latifolium

ONAGRACEAE

The deeper that sorrow carves into your being, the more joy you can contain.

__Kahlil Gibran

Botanical Description

Perennial plants grow up to 20 inches tall. Erect or sprawling, they can be somewhat woody at their base. Showy, 1-1 1/2 inch rose-purple (rarely white) flowers on short racemes. 4 petals, 4 sepals, 8 anthers. Lower stem leaves are alternate and broadly lanceolate.

Range

Grows in both eastern and western Washington and throughout the Pacific Northwest, south to Colorado and the Sierras. More common in the north part of the region, especially in Alaska and northern British Columbia.

Habitat and Ecology

In early summer streams and torrents of snowmelt tumble down mountain slopes, rearranging talus, scouring the thin soils. Life's tenuous footholds are often obliterated. In some of Cascadia's more northern ranges, as runoff diminishes, rivers of magenta begin to fill the water-carved courses. Here we witness River Beauty in its most vibrant strongholds.

Just as Fireweed leads life's re-entry after fire and other destruction, its close relative, River Beauty, can be the first pioneer following floods. Less common than Fireweed, River Beauty can be encountered in subalpine, alpine, and arctic areas, putting on glorious displays in the gravel bars of streams and rivers or along their sandy, stony banks. Plants grow along talus slopes near snow fields or springs where ample moisture persists for most of the summer. River Beauty recovers water-scrubbed land, drawing small animals and birds with food and shelter, its old leaves and stems contributing to soil formation.

Gesture

Growing low, sometimes sprawling into mats, sometimes erect, River Beauty plants are much shorter than Fireweed. But their flowers are larger, capturing attention with their showy color. The plants' reddish stem tips form short floral spikes, or racemes, bearing three to twelve flowers. Flower stalks emerge from the uppermost leaf axils. Floral color varies from red-violet, magenta, or bright pink to purplish-rose. The four petals, oval with pointed tips, are the same color as the four narrow sepals. Eight stamens and a pistil with a four-lobed stigma are prominent in the center of the wide-open blossom. Positioned horizontally on their racemes, blossoms are illuminated in the north country's extended mornings and evenings, when the low-slanting summer sun creates magic light.

The flower's ovary is inferior, that is, it grows below where the petals join the stalk. After fertilization the ovary becomes a long slender seed pod that will release lavish amounts of tawny-tufted seeds to ride the winds. Since River Beauty does not spread from rhizomes like Fireweed, it relies solely on these legions of wind-riders for propagation.

The leafy stems of River Beauty can be somewhat woody at their bases. Though *latifolium* means "broad leaf", and leaves are often widely oval, they can also be more lance-shaped. Smooth and fleshy, leaves have a bluish-green sheen from the fine waxy, or glaucous, coating that preserves moisture and shields them from extreme sunlight or cold. The tough, substantial qualities of their leaves and stems protect River Beauty plants in their high elevation and arctic habitats.

Other common names for *Chamerion latifolium* include Broad-leaf Willow-herb, Red Willow-herb, Alpine Fireweed, and Broad-leaved Fireweed. River Beauty and Fireweed (see also) are of the same genus, and are similar in many ways. For herbal and culinary purposes, the two species are interchangeable. The fresh flowers and leaves of both plants are mucilaginous, so they soothe inflamed and irritated tissue. They can be made into a poultice to ease sunburn and rashes. These parts can also be decocted into a tea for relief of cough spasms and asthma. Whole plants have laxative and astringent properties as well. Tender young leaves and shoots are tasty cooked or in salads.[1]

As flower essences, both River Beauty and Fireweed assist recovery from shock and trauma. River Beauty essence targets emotional recovery. It eases emotional pain after loss, allowing grief to be expressed. It helps one let go of what has been lost or what no longer serves, recognizing the potential for purification and growth in adversity.[2]

Medicine Story

Like Fireweed, River Beauty plays an important role in habitat renewal. While Fireweed appears after fire, avalanche, and human destruction, River Beauty follows water courses, erosion, and flood damage. It grows in environments with abundant moisture. Echoing its role in land restoration, River Beauty's flower essence is restorative to humans engulfed by emotional pain after trauma and loss. The plant's physical components soothe the body's trauma, relieving inflammation and quieting bronchial spasms. Oriental Medicine understands lung conditions, such as asthma and bronchitis, as reflecting unexpressed sadness or grief. On many levels we discover connections, as the various qualities of River Beauty express different octaves of the same essence.

Floods and streams swollen with spring runoff bring radical change. They dislodge boulders and trees, even send sides of mountains plummeting. Landscapes are altered, life-giving foundations swept away. These situations have their counterpart in human experience. Periodically, rapid change sweeps through our lives. We may be flooded with previously unexpressed emotions. Emotions correspond to the water element, for

---

1   Tilford
2   Johnson

they reflect the patterns and movements of water in all of its forms. Also, emotional expression releases the body's cleansing and healing fluids.

The flood waters of nature's cycles are purifying and releasing, part of her continual transformative processes. As areas are disassembled or washed away, life is offered new possibilities for reconstitution. Not only is River Beauty an affirmation of life's persistent resurgence, it is a living testimony to the beauty that can follow destruction. River Beauty opens the way for life's re-establishment, anchoring, attracting, and offering sustenance to other beings. At the same time, the rosy radiance of its flowers emanates a joyful, life-giving quality to its surroundings. The blossoms speak the language of inspiration and renewal. Their four-part structures point to the four directions of Mother Earth with her four elements that form, hold, and sustain us. (See Chapter 3.) Like the mother substance, River Beauty offers a matrix from which new life can develop. Its blossoms glow with rose and red-violet, colors of high frequency. At the edge of the visible spectrum, these colors uplift, quickening transformation.

In these times of ever-accelerated change, old foundations are being swept away on personal and global levels. Intense emotions are stirred, and collectively we attempt to block them, becoming rigid and unfeeling or engulfed and overwhelmed. Releasing the dams, allowing suffering to flow through us, assists the cleansing and revitalizing process. River Beauty medicine reminds us to surrender to the currents of change and to our human responses. It points to the new possibilities for joyful expression that emerge from life-altering events. River Beauty's streams of magenta blossoms trace the scars and clefts of the water-torn mountainsides. It offers us this vision, a vision which has the power to anchor and nourish new life.

Scotch Bluebell
Campanula rotundifolia
CAMPANULACEAE

It strikes me that what means most
Are the small things, begging
For my attention.

_Dian Greenwood

> **Botanical Description**
>
> 5 fused petals, blue to light purple, form radially symmetrical bell-shaped flowers. 5 green pointed sepals. Single flower or loose cluster of 2-15 on thin, wiry stalks at top of stem. Stem leaves linear, sessile, alternate; basal leaves are oval to heart-shaped with long petioles, and usually wither by time of blooming.
>
> **Range**
>
> Circumboreal.

Habitat and Ecology

Throughout the temperate regions of the Northern Hemisphere, Scotch Bluebell is common and widespread. It has adapted to considerably diverse environments. Plants grow in greatest abundance in the more northern parts of Europe, Asia, and North America, yet one can find them through most of Europe and the United States, except in the extreme southeastern states. In springtime this species spreads its charms through the dry grasslands and heaths of Scotland, where its flowers have so captured the hearts of the people as to become a national emblem. These are the Bluebells celebrated in the poetry of Robert Burns.

In Cascadia Scotch Bluebell grows from sea level to subalpine and even alpine areas. Thriving in dry to moist soils, these plants grace hillsides, valleys, prairies, open forests, stream sides, mountain meadows, cliffs, talus slopes, and crevices. They are late bloomers, their delicate bells appearing after the main floral show has peaked. After the colorful displays of Camas and Puget Balsamroot have faded from the prairies south of Puget Sound, Scotch Bluebell gives us something to look forward to. In late spring and early summer its flowers' gentle beauty accompanies the song of the Swainson's Thrush. Along the cliffs and slopes of the west Columbia Gorge, it is a mid-summer flower, while late summer finds

Scotch Bluebell in high mountain meadows blooming with Asters and Explorer's Gentian. At Mt. Rainier the lavender-blue flowers may appear in a tangle of dried meadow foliage or at the foot of a glacier. They may still surprise us as the high country begins to blaze with autumn color.

This dainty, flimsy-looking plant survives the fierce storms prevalent in high, exposed, and more northern habitats. Its thin, delicate foliage is actually an asset, enabling plants to bend and yield in violent rain and wind. The minimal surface area of the foliage is well-adapted to extreme dryness and cold. Also, the leaves and stems contain bitter alkaloids that repel foragers.[1] Scotch Bluebell has therefore remained abundant in the British Isles where domestic sheep and goats abound. Similarly, in the mountains of the North American west, plants are safe from these animals' less rapacious wild relatives.

Gesture

The growth habits and characteristics of Scotch Bluebell vary considerably with the bounties and rigors of its diverse habitats. When soil is deep, rich, and moist, plants may grow up to two feet tall and bear abundant flowers. In alpine zones, two to five inch plants often have but one flower per stem. As with many high elevation flowers, this bloom is usually large, dwarfing the rest of the plant. Plants amass in low clumps that can be laden with blossoms. While foliage in moist, nutrient-rich conditions tends to be smooth, in dry and alpine areas, leaves are shorter and covered with fine, moisture-preserving hairs.

In all of its forms and habitats, Scotch Bluebell is slender and delicate, with thin, often drooping stems, thread-like flower stalks, and narrow to grass-like upper leaves. While *Campanula*, meaning "little bell", describes the dainty flowers, one might wonder at the species name, *rotundifolia*, or "round-leaved". The plant grows round or heart-shaped basal leaves,

---

1   Clark

but they are often completely withered by bloom time; therefore, the linear stem leaves are more noticeable. The two types of leaves sharply contrast. The few wide basal leaves grow from long, thin petioles, while the more plentiful, narrow upper leaves alternate stalkless up the stems. Stems and leaves have a milky sap.

Everything about the flower is soft, gentle, and dainty—in a word, charming. The five fused petals are washed with light blue, lavender, or periwinkle. While buds begin erect, thanks to the delicacy of their petioles, flowers nod downwards as they open. This orientation, which adds to their bell-like gesture, protects the flowers' pollen and nectar from the rain.[2] On their delicate flower stalks, the little bells tremble in a mere whisper of breeze. The style extends the length of the corolla, and, tipped with a three-lobed stigma, it becomes a perfect "clapper" for the "bell". The filaments of the five stamens are shorter than the style; wider at their bases, they curl above the inferior ovary and cover the nectaries. The size and position of reproductive parts and nectaries, along with synchronized anther and stigma-ripening, conspire with bees' hairy legs in a sophisticated pollination strategy.[3]

Scotch Bluebell, so slender and dainty in its above-ground appearance, has a well-developed root system, including a taproot and slender rhizomes.

Growing across many lands, *Campanula rotundifolia* has an abundance of common names. Its English names include, Bluebells-of-Scotland, Bellflower, Common Harebell, and Scots Bluebells. From the Haida language its name translates as "blue rain flowers". Children were told that picking these flowers would bring rain.[4]

In Scotland a flower essence is made from the blossoms of this species. The essence is for imparting a sense of prosperity and faith, enabling

---

2   *Ibid.*

3   *Ibid.* See p.505 for a more detailed explanation.

4   Pojar et al.

one to release possessiveness and material concerns. It assists one to let go of limitations, and to receive gentleness, strength, and clarity, while reclaiming wildness and kinship with nature.[5]

Medicine Story

With Scotch Bluebell all things fragile and flimsy play a protective role, enabling survival in harsh circumstances. The slender stems are pliable; yielding, they are unbroken by gales and downpours. Because its petiole is dainty, the flower bends toward the ground, sheltering its pollen from the rain. Grass-thin stem leaves, with minimal surface area, are well-suited for extreme conditions. Hidden from view is the plant's ability to repel predators and to anchor itself with strong, deep roots. Scotch Bluebell is a very hardy plant. It flourishes in many habitats, responding and adjusting to a wide range of challenges. Behind the subtle, almost ethereal color and diminutive charm of its flowers is unsuspected strength.

Strength through yielding and responsiveness is part of the essential nature of Scotch Bluebell. The nodding flowers speak of humility, attunement to the Earth, attentiveness to the breeze. In our imaginations they are bells, little paper lanterns; they are open, spacious, light. Their gesture starkly contrasts with the nearby Explorer's Gentian blossoms, boldly colored, reaching skyward from erect and sturdy stems. The soft Bluebell could be overlooked, yet we find ourselves leaning closer to catch its whisper. What gentle music are these flowers playing? Or are they demonstrating—and eliciting from us—the capacity to *listen*, to cup our ears and become empty? Bells punctuate the moment, gather our awareness. They gather people in celebration, sorrow, and worship. Ringing bells purify space, clear attention, awaken us to the sacredness of *now*. Close your eyes with Scotch Bluebell's flowers—listen. Previously unheard sounds come into focus. We become aware of how much eludes

---

5   Harvey, Cochrane

our awareness. Our inner ear attunes, and the silence within begins to speak. We meditate with petals shaped like ears of animals; inside is iridescence, the shimmer of soft hairs.

In this era, problems of unprecedented magnitude threaten to overwhelm us. As we resist them they become stronger. Scotch Bluebell instructs us in the medicine of yielding, of response born of presence and attentiveness. This fragile-looking little plant, whose nature is to bend and tremble, teaches the power of humility. We are conditioned to amass force to meet adversity, only to augment and become part of that adversity. What happens if we shift our attention to soft lavender-blue flowers? There is a whole subtle realm around us whispering. How better to meet the fierce danger of life than to deeply *listen*? ...to listen to other people, to the voices of nature, to the one clear note that set your life in motion. With this humble gesture, we give to the world all it ever asked of us.

Explorer's Gentian

Gentiana calycosa

GENTIANACEAE

And the time came when the risk to remain tight in a bud was more painful than the risk it took to blossom.

—Anais Nin

Botanical Description

A low growing perennial with two to many unbranched stems 4-10 inches long. Leaves are smooth, nearly round, and grow directly from the stem in opposite pairs. A set of flowers emanates from the top of the stems, sometimes also on the upper leaf axils. Flowers are 1-2 inches long, trumpet-shaped, parts in 5's, except for a single pistil; sepals are fused. Petals are deep blue to blue violet with green spots. Petals are united at base and flare at tips with fringed or fine-toothed "pleats" between.

Range

Grows from 4000-13,000 feet from British Columbia south into northern California and east into the Rocky Mountain states. It is therefore found in the Olympic, Cascade, Sierra, and the Rocky Mountain ranges.

Habitat and Ecology

The flowers of Explorer's Gentian are the indigo dream of summer's passing. They are just beginning to bloom as Huckleberry leaves redden, when Scotch Bluebells, Pearly Everlastings, and late Asters are yet found amidst Lupine pods and the brown seed heads of Heather and Pedicularis. Nestled in meadows golden with autumn, the intense blue of this Gentian almost startles.

Explorer's Gentian is widely distributed throughout the northern portion of the west at elevations of 4,000 feet and above. Whether in the coastal or inland mountains, this species grows where the land retains abundant wetness in late season. In moist to boggy meadows, its trumpets can surprise us; here the jeweled presence of summer flares one last time.

Gesture

The species' name *calycosa* means cup-like, a description that applies to the large tubular flowers. However, the midnight blue buds can remain tightly closed for weeks, and flowers are quick to shut with cloudy conditions or at day's end. Only in bright sunlight do the five petals unwrap like a pinwheel; petal tips flex backwards, and flowers become long fluted cups, extending heavenward. At these times the sky seems to pour into them its crystal blue, and they shine like porcelain.

The almost round leaves are succulent, growing horizontally in opposite pairs that shift directions up the red, wiry stems. As a prelude to the terminal flower clusters, near the top, leaves begin to reach upward in a cup-like gesture, soon becoming the calyx that enfolds the growing buds. Stems and plants grow in clusters, their fleshy leaves a vigorous display of shiny green. The plants are the remembrance of the season's blessings of snow and rain, a last luxuriant flourish when other plants—even blossoming ones—are dry and sparse. This Gentian's ability to hold water and color is its gift in this time of transition.

Explorer's Gentian has a close relationship with water and also with the sky. Its sturdy stems grow straight up, and elongated flower tubes reach for the heavens. Only on stream banks do they sometimes curve down towards the water. Proceeding up the stem, the leaves begin an open embrace skyward that culminates in the floral chalice. The sharply pointed buds as well as the relative largeness (as compared to the rest of the plant) of the flowers with their vivid sky colors also reveal the plant's celestial affinity.

The flower's color is an interplay of blues, deep violets, and white. The outsides of the petals are brushed with lines of purple that bleed into blue. Reflexed petal tips are vivid blue, while inside the fused petals, white is dappled with blue. Flecks of olive green are sprinkled on petal tips and form delicate lines in the light interior.

The exterior coloration defines separate petals, yet they are fused from the base most of the way up, separating only at the tips. Stamens are separate at the base, then clasp near the anthers and enfold the pistil, echoing the petals in reverse. The blue pleats between petal tips split into fingers of fringe, perhaps ideal for dusting furry bees as they disappear into the deep throat of the flower. The poles of fusion and differentiation dance in this flower. Noticeable are the tenacity of the dark, closed buds and the flowers' hair trigger closures at fading light. Yet, with the magic touch of sunlight, petals individuate, spiraling open.

Medicine Story

Just when we are resigned to the end of flower season, the Explorer's Gentian offers yet another drink from sparkling skies and deep mountain lakes. This shockingly blue flower arrests us in the present moment, offering timeless renewal. The plant carries verdant power through summer's decline. Sit with Explorer's Gentian to absorb its deep and rich medicine. It calls us to look up, to be filled with the sky.

The unique beauty of this Gentian is all the more precious because it graces the end of the growing season. We often believe that it is too late to actualize our unique dreams and potential. Conditioning tells us that opportunities and other people have passed us by; burdens and obstacles seduce us into giving up, or into never trying in the first place. Explorer's Gentian meets our reluctance and resignation and teaches us to allow light to enter our souls. Its beautiful color inspires and opens us to the spiritual aspects of life. The radiance of this flower encourages us, at last, to flower—to overcome discouragement through awareness of the beauty and perfect timing of our contribution.

Glossary

alkaloid: a toxic water-insoluble compound containing nitrogen. Many plants are protected from animal predation through the production of alkaloids.

anther: a pollen-producing sack located at the stamen's tip.

axil: the angle created where the leaf joins the stem.

banner: the upper, usually largest, petal of a flower in the Pea (Leguminosae) family.

basal leaf: a leaf that grows at or near the ground at the base of a stem.

bilaterally symmetrical: describes an irregular flower whose left and right sides are mirror images of each other. The symmetry found in an animal or human face or body.

bilobed: divided into two lobes.

bloom: a white powdery or waxy substance coating the surfaces of some leaves and berries.

bract: a leaf-like structure that can be part of an inflorescence.

bulb: an underground fleshy structure that stores water and nutrients for the plant.

calyx: the whorl of sepals that encloses the rest of the flower.

circumboreal: belonging to the band of northern latitudes circling the entire globe.

composite flower: an inflorescence consisting of many separate flowers tightly packed together.

compound leaf: a leaf comprised of separate leaflets or dissected one or more times into lobes.

corm: a thick, vertical underground stem that stores nutrients; usually has a papery covering.

corolla: all of a flower's petals, referred to collectively.

disk flower: one of the small, tube-shaped flowers in the inflorescence of many composite flowers; often comprises the center, but sometimes the whole, of a composite.

endemic: growing only in a particular region.

entire: having smooth margins without teeth or lobes.

erect: growing upright from the ground.

filament: a stamen's stalk, bearing the anther at its tip.

floret: a small individual flower that comprises a larger inflorescence.

glabrous: smooth, without hairs.

glaucous: coated with a waxy or powdery film (often referred to as bloom), which gives a whitish or bluish cast, for example, to fruits or leaves.

habitat: an environment with a particular set of conditions making it a suitable home for a given plant or other living creature.

herbaceous: refers to a plant or plant material that is not woody.

hybrid: a plant that has resulted from the breeding of two different species.

inferior ovary: an ovary positioned below the convergence of the stamens, petals, and sepals.

inflorescence: an entire cluster of flowers.

involucre: a cluster of bracts beneath the head of a composite flower.

keel: the lower two petals of a Pea (Leguminosae) flower. Joined, they form a shape like the keel of a boat and they enclose the flower's reproductive parts.

lanceolate: refers to a leaf or petal shaped like a blade or spear; that is, wide at the base, elongated, and tapered to a sharp tip.

leaflet: a leaf-like part of a compound leaf.

linear: long and narrow with straight, or nearly straight, sides; needle-like.

lithosol: means "rock-soil", and refers to a dry habitat characterized by crumbly rock and a very thin soil layer. Plants here tend to grow close to the ground, often in mats, to protect themselves from the sun's heat and moisture-leaching winds.

lobed: refers to a leaf or petal whose margins have deep, round indentations.

mycorrhiza: a tiny, highly efficient nutrient-conducting structure formed by plant roots and fungi, creating a mutually beneficial network.

nectar: a sweet, pollinator-attracting liquid secreted by many flowers.

nectary: a floral structure that produces nectar.

ovary: the seed-bearing structure at the base of the pistil. Matures into the fruit.

ovate: egg-shaped.

palmate: shaped like the palm of a hand, fingers extended; that is, with parts radiating from a central point. Used to describe a pattern of veins, leaf lobes, or leaflets in a compound leaf. In their overall shape, deciduous, or broadleaf, trees mainly follow this pattern, in contrast to the generally **pinnate** (see also) shape of conifers.

panicle: a branched flower cluster in which flowers open successively from the bottom up.

pedicel: the stalk of a flower or fruit.

perianth: a collective term for the calyx and corolla, that is, the sepals and petals combined.

petiole: a leaf stalk.

pinnate: a pattern of growth, which describes leaf veins and leaflets in a compound leaf. In this pattern there is a single central axis from which secondary branches emerge laterally. This pattern imitates the general growth of a pine tree or other conifer, and contrasts with a **palmate** (see also) pattern of growth.

pistil: the female structure at the very center of a flower, comprised of the ovary, style, and stigma.

prostrate: trailing or growing more or less on the ground.

pubescent: hairy or woolly.

raceme: a flower cluster in which the individual flower stalks grow from an unbranched stem. Flowers mature from base to tip.

radially symmetrical: describes a regular flower, that is, one whose parts are similar in shape and size and radiate from a central point.

ray flower: one of the elongated florets of a composite flower, usually making up its outer circle, but sometimes the whole flower. Appearing like a single petal, each is actually 5 fused petals.

rhizome: an underground stem or rootstalk which sprouts new stems, assisting the plant to spread in a creeping manner. Some rhizomes are thick and starchy, storing nutrients for the plant.

reflexed: bent or turned backward or downward.

rosette: a basal whorl of leaves growing from the root crown.

scree: small gravel-like rock debris.

sepal: a modified leaf that encloses the bud and makes up the outermost part of a flower.

serpentine: a type of rock with high concentrations of magnesium, iron, nickel, and toxic metals.

sessile: without a stalk or petiole.

shrub-steppe: an open arid habitat with relatively deep soils characterized by sagebrush, bunchgrass, and scattered shrubs.

solitary: single-flowered, as opposed to flowers growing in clusters.

stamen: the male structure of a flower consisting of a filament, or thin stalk, and pollen-bearing sack, or anther. Typically, several stamens surround the pistil, or female structure located at the flower's center.

stigma: the sticky tip of the pistil where the pollen is received.

style: the pistil's stalk, which transports male reproductive cells from the stigma to the ovary.

superior ovary: an ovary positioned above the petals.

talus: rock fragments or boulders at a cliff base or on a slope, less fine than scree.

taproot: a plant's large central root, usually growing straight down, deep into the ground.

tepal: a petal or sepal of a flower in which the petals and sepals are identical in shape and color.

terminal: growing from the tip or end.

umbell: a circular floral cluster in which all the flower stalks radiate from the same point on the stem, forming an umbrella-like structure.

whorl: three or more leaves, sepals, or petals growing from the same point around a stem.

References

Adams, George and Whicher, Olive. *The Plant Between Sun and Earth.* Boulder: Shambhala Publications, Inc., 1982.

Appelgren, Barbara June. "Awareness Through Vision Dr. Harry Sirota". http://www.lclark.edu/~miller/sirota.html

Barnard, Julian. *Bach Flower Remedies Form & Function.* Hereford, Great Britain: Flower Remedy Programme, 2002.

Biek, David. *Flora of Mount Rainier National Park.* Corvallis: Oregon State University Press, 2000.

Blackwell, Laird R. *Wildflowers of the Sierra Nevada and the Central Valley.* Vancouver, BC: Lone Pine Publishing, 1999.

Carville, Julie Stauffer. *Hiking Tahoe's Wildflower Trails.* Vancouver, BC: Lone Pine Publishing, 1989.

Clark, Lewis J., Trelawny, John G., ed. *Wildflowers of the Pacific Northwest.* Madeira Park, BC Canada: Harbour Publishing, 1998.

Craighead, John J., Craighead, Jr., Frank C., and Davis, Ray J. *A Field Guide to Rocky Mountain Wildflowers: Northern Arizona and New Mexico to British Columbia.* Boston: Houghton Mifflin Company, 1991.

Curtis, Susan and Fraser, Romy. *Natural Healing for Women: Caring for Yourself with Herbs, Homoeopathy and Essential Oils*. London: Pandora Press, 1991.

Eiseley, Loren. *How Flowers Changed the World*. New York: Random House, Inc., 1996.

Elpel, Thomas J. *Botany in a Day: Herbal Field Guide to Plant Families, 4th edition*. Pony, Montana: HOPS Press, 2000.

Greene, Jennifer. "Water Wonders". Bioneers Conference Workshop. San Rafael, CA, October 2004.

Grohmann, Gerbert. *The Plant, vol. 2*. Kimberton, PA: Bio Dynamic Literature, 1989.

Hall, Alan. *Water, Electricity and Health*. Lansdown: Hawthorn Press, 1998.

Harvey, Clare G. and Cochrane, Amanda. *The Encyclopaedia of Flower Remedies*. London: Thorsons, 1995.

Jay, Tom and Matsen, Brad. *Reaching Home: Pacific Salmon, Pacific People*. Anchorage: Alaska Northwest Books, 1995.

Johnson, Steve. *The Essence of Healing: A Guide to the Alaskan Essences, 2nd edition*. Homer, Alaska: Alaskan Flower Essence Project, 2000.

Jolley, Russ. *Wildflowers of the Columbia Gorge: A Comprehensive Field Guide*. Portland: Oregon Historical Society Press, 1988.

Jung, C. G. *Modern Man in Search of a Soul*. New York: Harcourt, Brace & World, Inc., 1933.

Kaminski, Patricia and Katz, Richard. *Flower Essence Repertory*. Nevada City, CA, 1994.

Kaplan, Robert-Michael. *The Power Behind Your Eyes.* Rochester, Vermont: Healing Arts Press, 1995.

Krafel, Paul. *Seeing Nature: Deliberate Encounters with the Visible World.* White River Junction, VT: Chelsea Green Publishing, 1999.

Kreisberg, Joel. *Apiaceae: The Carrot Family: A Homeopathic Field Guide.* Berkeley, CA: Teleosis Homeopathic Publishing, 2000.

LaDuke, Winona. "Indigenous Mind". http://www.resurgence.org/resurgence/articles/laduke.htm

Liberman, Jacob. *Light: Medicine of the Future.* Santa Fe, N.M.: Bear & Company, 1991.

Mathews, Daniel. *Cascade-Olympic Natural History: A Trailside Reference.* Raven Editions in conjunction with the Portland Audubon Society, 1990.

McIntyre, Anne. *Flower Power.* New York: Henry Holt and Company, 1996.

Moore, Michael. *Medicinal Plants of the Pacific West.* Santa Fe, NM: Red Crane Books, 1993.

Nature Conservancy. "The Prairies & Oak Woodlands of South Puget Sound". Information Pamphlet.

Niehaus, Theodore F. and Ripper, Charles L. *A Field Guide to Pacific States Wildflowers: Washington, Oregon, California and adjacent areas.* Boston: Houghton Mifflin Company, 1976.

Pogacnik, Marko. *Healing the Heart of the Earth: Restoring the Subtle Levels of Life.* Forres, Scotland: Findhorn Press, 1997.

Pojar, Jim and MacKinnon, Andy, et al. *Plants of the Pacific Northwest Coast: Washington, Oregon, British Columbia, & Alaska.* Vancouver, British Columbia: Lone Pine Publishing, 1994.

Rathbun-Holstein, Melissa. "Native Orchids of Washington". Native Plant Society Talk. Olympia, WA, 9 October 2006.

Ross, Robert A. and Chambers, Henrietta L. *Wildflowers of the Western Cascades.* Portland, Oregon: Timber Press, 2000.

Ryrie, Charlie. *The Healing Energies of Water.* Boston: Journey Editions, 1999.

Schneider, Michael S. *A Beginner's Guide to Constructing the Universe: The Mathematical Archetypes of Nature, Art, and Science, A Voyage From 1 to 10.* New York: HarperPerennial, 1995.

Schwenk, Theodor and Schwenk, Wolfram. *Water, the Element of Life.* New York: Anthroposophic Press, 1989.

Shebitz, Daniela. "Incorporating Traditional Land Management into Restoration of Anthropogenically-Maintained Beargrass Habitat on the Olympic Peninsula Lowlands". Native Plant Society Talk. Olympia, WA, 21 November 2005.

Stokes, Donald and Lillian. *A Guide to Enjoying Wildflowers.* Boston: Little, Brown and Company, 1985.

Strickler, Dee. *Northwest Penstemons: 80 Species of Penstemon Native to the Pacific Northwest.* Columbia Falls, Montana: Flower Press, 1997.

Swimme, Brian. *Canticle to the Cosmos.* Audiotape. Boulder, CO: Sounds True Inc., 1995.

Taylor, Ronald J. and Douglas, George W. *Mountain Plants of the Pacific Northwest*. Missoula, Montana: Mountain Press Publishing Company, 1995.

Thich Nhat Hanh. "The Present Moment" in *Voices of Wisdom*. Audiotape. Boulder, CO: Sounds True Inc., 2000.

Tilford, Gregory L. *From Earth to Herbalist: An Earth-Conscious Guide to Medicinal Plants*. Missoula, Montana: Mountain Press Publishing Company, 1998.

Tolle, Eckhart. *A New Earth: Awakening to Your Life's Purpose*. New York: Plume, a member of Penguin Group (USA) Inc., 2005.

Turner, Mark and Gustafson, Phyllis. *Wildflowers of the Pacific Northwest*. Portland, OR: Timber Press, 2006.

Visalli, Dana. "Penstemons: The Mystery of the Missing Anther" in *The Methow Naturalist*. Spring 2006 V11 N2.

Zajonc, Arthur. *Catching the Light: The Entwined History of Light and Mind*. New York: Oxford University Press, 1993.

The Author

Julia Brayshaw, MA/ABS is a lifetime devotee of nature's beauty. A depth psychotherapist for twenty years, she is seeking to integrate these twin callings. She is a flower essence practitioner, certified by the Flower Essence Society of California. Julia is an avid wilderness explorer, and has hiked and backpacked extensively in North America. In the past decade she has also explored the British Isles, Brazil, Greece, and Norway, encountering new bioregions and new plant communities. Her passion is to honor *place,* to join others in opening to wonder and respectful relationship.

The Artist

Karen Lohmann is an artist and healer. Her interest in exploring the natural world has led her from designing with flowers and plants to the interior work of Hospice care and Flower Essence Therapies. With pastel art, she continues the dialogue, reflecting color and shape on these life mysteries. She is pleased to fulfill a dream of creating a deck of flower cards that encourages interaction with the soul of nature and is thrilled to be a collaborator with Julia Brayshaw in *Medicine of Place*. Karen is a Comfort Care Therapist, working as a Certified (Flower Essence Society) Flower Essence Practitioner. She creates fresh floral designs for FlorAbunda. Her chalk pastels have shown in City of Olympia sponsored Arts Walks and Studio Tours. Her Public Art Projects as a Landscape Designer include: the "Seven Oars Parks" 1 and 2, "Stone Amphitheater" Earthworks at Bigelow Park and the Delbert McBride Ethnobotanical Gardens at The Washington State Capitol Museum. Karen has lived in Kyoto, Japan—studying Zen, traditional Japanese Landscaping, and Ikebana—and in Salzburg, Austria. She has hiked in the Lakes District of England, the Irish Dingle Peninsula, and trekked in Nepal, pursuing a love of hosteling. Karen and spouse, Joe Tougas, have 3 grown children, and a cute grandson.